BASIC ARITHMETIC
Second Edition

BASIC ARITHMETIC

Second Edition

Ross F. Brown

Fanshawe College of Applied Arts and Technology

HarperCollins*Publishers*

ISBN 0-673-18017-4

78910-WEB-939291

Preface

The second edition of *Basic Arithmetic* is designed to develop computational skill along with an understanding of basic arithmetic operations. In addition to the basic topics of arithmetic, there is an introduction to the metric system and an introduction to some basic ideas of algebra and plane geometry.

No mathematical prerequisite is assumed for students using this text. Moreover, students with weak reading skills can also experience success. Classes or students who can work with whole numbers may wish to begin with Unit 5 on multiplying and dividing fractions.

Typical subsequent courses are beginning (introductory) algebra, technical mathematics, general education mathematics, mathematics for data processing, or mathematics for business, secretarial science, and related areas. This book can be used in traditional lecture classes, for individualized study, or for a combination of the two learning situations.

The first edition was developed during many years of teaching students in a college retraining system. That edition was used by many students having a wide variety of skill and reading levels. Many of these students have gone on to complete further course work in mathematics. The second edition represents changes requested by the many users of the first edition.

Content Features

- An adult approach to the coverage of **fractions, decimals,** and **signed numbers** helps students understand these traditionally difficult topics.

- Many more **word problems** give students practice solving realistic problems.

- Coverage of the **metric system** now involves learning metric approximations. This relieves students from lengthy and tedious calculations.

- Expanded coverage of **geometry** includes more on angles, straight lines, and parallel lines and less on volume.

- The chapter on **combined operations** now immediately precedes Unit 16 on preparing for algebra, serving as a logical introduction to that material.

- Additional coverage of **signed numbers** provides the necessary background for further courses in mathematics.

Pedagogical Features

- The **simple, uncluttered** design makes the book easy to use and visually appealing.

- A **second color** has been used more extensively to illustrate and explain procedures and to emphasize important rules.

- An **easy-to-read typeface** facilitates students' understanding of the material.

- **Examples** are plentiful, clear, and easy to follow. They prepare students to work the exercise sets.

- The 2800 **exercises** in the text provide students with ample practice. Answers (given upside down) follow each exercise set to provide students with immediate feedback and reinforcement.

- The **Instructor's Guide** includes supplementary exercises for a few difficult areas, a diagnostic pretest, and five unit tests (with answers) per unit.

- Adult-oriented **illustrations** help students visualize techniques presented and concepts discussed in the text.

- An **index** to the text helps students locate topics for review during the course and afterward.

Acknowledgments

For their helpful comments on the manuscript, I thank Mary M. Blyth, Detroit College of Business; William Covell, Mount Hood Community College; Charles A. Bower, St. Philip's College; Patricia L. Hirschy, Delaware Technical and Community College; Jewell Hodge, Richland College; Anne Martin, Delta College; Judy Mee, Oklahoma City Community College; Donald Reichman, Mercer County Community College; Jack W. Rotman, Lansing Community College; and Alfred F. Soprano, Clark County Community College.

As a fellow teacher, I appreciate the value of a second opinion. Therefore, I welcome your comments, criticisms, and/or brief evaluations of the book.

Ross Brown
c/o College Mathematics Group
Scott, Foresman and Company
Glenview IL 60025

Contents

1 Adding Whole Numbers

Aims you toward:

- Adding numbers and getting the right answer.

- Naming numbers.

- Knowing the place value for ones, tens, hundreds, and thousands.

In this unit you will work exercises like these:

- Add 298 + 41 + 3.

- What is the number 321,406,327 in words?

- What is the value of the middle "2" in 222?

Let's go . . . !

Adding Numbers

This unit starts off with a 1, which seems like a good beginning when you COUNT. We could use the digit 1 to stand for all our numbers:

$1 + 1$ is 2 $1 + 1 + 1$ is 3 $1 + 1 + 1 + 1$ is 4 $1 + 1 + 1 + 1 + 1$ is 5   and so on.

But that gets clumsy pretty fast. It is easier and takes less space if we use 0, 1, 2, 3, 4, 5, 6, 7, 8, 9. All our numbers are shown by combinations of these ten digits.

That first digit, 0, or ZERO, comes up a lot when you add. The nice thing about zero is that any number plus zero is the number you started with.

Example   Take 3 and add 0 to it.

(addend)   3   Stack it up the other way, and it still   0
(addend) $+0$ adds up to the same answer. ⟶ $+3$
(sum)   3   3

What about adding other numbers? Let's see.

Example   Find the sum of $21 + 7$. (When you see $+$ you know you should ADD.)

21   Line up the numbers on the right-hand side.
$+ 7$

21   Begin by adding the right-hand COLUMN. $1 + 7$ is 8.
$+ 7$
  8

21   Go left to the next column. $2 + 0$ is 2.
$+ 7$
 28   Keep the answer lined up, too.

Examples

7	40	21	46	457	3,612
$+22$	$+ 3$	$+10$	$+13$	$+232$	$+7,215$
29	43	31	59	689	10,827

To Add	Line up the numbers on the right-hand side. Add down the first column on the right. Move left. Add down the columns as you come to them. Keep the answer lined up. To check your work, try adding from the bottom up. You should get the same answer both ways.

Exercises

First try some that are already lined up in columns. Make sure your answer is lined up!

1.	7 +9	**2.**	8 +7	**3.**	7 +51	**4.**	51 + 8	**5.**	16 + 3	**6.**	42 +16

7.	6 +5	**8.**	7 +4	**9.**	3 +8	**10.**	13 + 5	**11.**	25 +33	**12.**	56 +41

13.	9 +7	**14.**	7 +8	**15.**	23 +32	**16.**	34 +43	**17.**	45 +54

18.	21 +12	**19.**	614 +735	**20.**	8,307 +5,041	**21.**	987 +412	**22.**	8,593 +6,403

23. A small cargo plane was licensed to carry 9,000 pounds. After loading 5,760 pounds on board, the pilot told the crew that they could add another 3,238 pounds safely. Was he right?

24. A community college has a part-time enrollment of 3,269 students. If the number of full-time students is 6,728, how many students are enrolled altogether?

More Exercises

Now you will need to line up the numbers in columns before you add.

1. 21 + 35   **2.** 40 + 59   **3.** 27 + 72   **4.** 7 + 9

5. 56 + 22   **6.** 11 + 38   **7.** 34 + 23   **8.** 10 + 87

9. Let's raise the   **10.** 400 + 59   **11.** 207 + 72   **12.** 700 + 200
ante.
Add 21 + 350.

$$
\begin{array}{r}
\downarrow\downarrow\downarrow\\
21\\
+\,350\\
\hline
371
\end{array}
$$

13. 2,700 + 190   **14.** 9,506 + 4,122   **15.** 634 + 1,234

16. Grandfather, who was 99 years old, said that the total of his grandchildren's ages was 10 years more than his own age. Margie was 16, and her brother Bart was 22. Jay was 18, and his sister Betty was 23. This left the twins, who were 15 years old. Was Grandfather right?

17. Anita and Sue spent 5 days hiking along the north shore of Lake Superior. The first day, they covered 26 miles. Over the next 3 days, they covered 12 miles, 14 miles, and 9 miles. On the last day, they managed 31 miles. How far did they hike?

18. On their next trip, Anita and Sue decided to travel by water. When moving from one lake to another, they carried their 76-pound canoe, along with two backpacks weighing 28 pounds and 32 pounds, 10 pounds of cooking utensils, and a 2-pound radio. How much weight were they carrying altogether?

Answers

17. 92 miles   **18.** 148 pounds

16. Yes, because the grandchildren's ages add up to 109, the same number you get by adding 10 to Grandfather's age.

11. 279   **12.** 900   **13.** 2,890   **14.** 13,628   **15.** 1,868

6. 49   **7.** 57   **8.** 97   **9.** 371   **10.** 459

1. 56   **2.** 99   **3.** 99   **4.** 16   **5.** 78

Some numbers take up more than one group.

Example   The moon is about 2,100 miles in diameter.

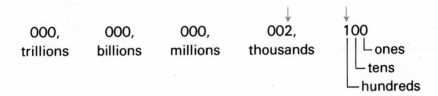

We read this number as TWO THOUSAND, ONE HUNDRED.

Example   Write 321,406,300 in words.

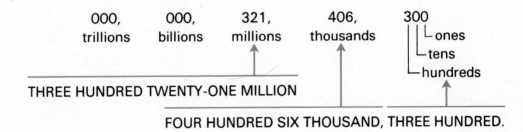

Example   The nearest star to the Earth is 25,000,000,000,000 miles away.

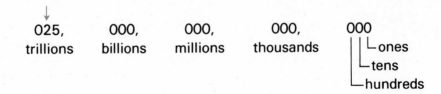

Since the 25 is in the "trillions" place, we name the entire number "TWENTY-FIVE TRILLION."

Now try some exercises in naming numbers.

Exercises

Write the following numbers in words. Use the string of zeros as a guide.

000, trillions	000, billions	000, millions	000, thousands	000

└ ones
└ tens
└ hundreds

1. 72 _____

2. 136 _____

3. 2,000 _____

4. 2,000,000 _____

5. 100,000 _____

6. 358,000,000 _____

7. 2,421 _____

_____

8. 25,000 _____

9. 276,641 _____

_____

10. 3,744,298 _____

_____

More Exercises

Now try putting the words into numbers.

000,	000,	000,	000,	000
trillions	billions	millions	thousands	└ ones └ tens └ hundreds

1. Three thousand, three hundred _____

2. One million, two hundred thirty thousand, six hundred _____

3. Five hundred twenty trillion _____

4. Two billion, one hundred ninety-three million _____

5. Seventy-eight million, six hundred twenty-one thousand, eight hundred forty-nine

_____

6. Seven hundred six thousand, seven hundred six _____

7. Forty-nine billion, one hundred _____

8. Fifty-four thousand, nine hundred eighty-five _____

_____

9. Two hundred twenty-four million, six hundred twelve thousand, one hundred

forty-six _____

10. One trillion, one billion, one million, one thousand

_____

Answers

9. 224,612,146   10. 1,001,001,001,000
5. 78,621,849   6. 706,706   7. 49,000,000,100   8. 54,985
1. 3,300   2. 1,230,600   3. 520,000,000,000,000   4. 2,193,000,000

Place Value

Here's where naming numbers connects with adding: PLACE VALUE. Take the number 111. Each "1" means something different because of where it is in the number. The "1" at the left means 100; the middle "1" means 10; the "1" at the end means 1. So 111 really means $100 + 10 + 1$, but the shortcut way to write it is 111.

Example   Suppose you bought a new color TV for $625. If you could only count by 1's, you would have one-dollar bills lined up end to end as long as a dozen classrooms. To avoid this, COUNT BY 10's AND BY 100's.

Look at the price of $625.
How many 100-dollar bills are there?
How many   10-dollar bills are left?
How many   1-dollar bills are left?

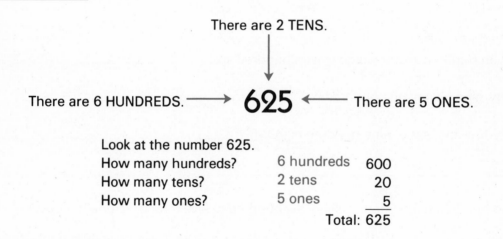

There are 2 TENS.

There are 6 HUNDREDS. ⟶ **625** ⟵ There are 5 ONES.

Look at the number 625.
How many hundreds?   6 hundreds   600
How many tens?   2 tens   20
How many ones?   5 ones   5
                          Total: 625

The 2 is in the tens place.

The 6 is in the hundreds place. ⟶ **625** ⟵ The 5 is in the ones place.

If you think 625 is a lot of money, just wait. We're going to add to it.

Place Value and Adding Numbers

Let's add $24 for a TV aerial to the $625.

```
  625
+  24
─────
  649
```

Say a TV guide costs another $1. Then

```
  625
   24
+   1
─────
  650
```

How's that? Let's take it again, in slow motion. Add 625 and 25:

```
  625
+  25
```
5 + 5 is 10, but you can't put 10 in the ONES PLACE. What you have is 0 ones and 1 ten.

```
  ¹
  625
+  25
─────
    0
```
Show this by putting 0 in the ones column. (As a reminder to yourself, put a small "1" over the tens column to mean 1 ten.)

```
  ¹
  625
+  25
─────
   50
```
Now add 1 + 2 + 2, meaning 1 ten + 2 tens + 2 tens. That's 5 tens. Put the "5" in the tens column.

```
  ¹
  625
+  25
─────
  650
```
Going to the left, you have 6 hundreds. 6 hundreds + 0 hundreds is 6 hundreds. Put the "6" in the hundreds column.

Let's zoom in on that one more time.

$$
\begin{array}{r}
1 \\
625 \\
+\ \ 25 \\
\hline
650
\end{array}
$$

← Write the 0 in the ones column.

Carry the 1 to the tens column.

Example   Add 625 and 26.

$$
\begin{array}{r}
625 \\
+\ 26 \\
\hline
\end{array}
$$
5 + 6 is 11. That's 1 one and 1 ten.

$$
\begin{array}{r}
^1 \\
625 \\
+\ 26 \\
\hline
1
\end{array}
$$
Put 1 in the ones column.
Carry the other 1 to the tens column.

$$
\begin{array}{r}
^1 \\
625 \\
+\ 26 \\
\hline
51
\end{array}
$$
Add 1 + 2 + 2 (tens). Put 5 in the tens column.

$$
\begin{array}{r}
625 \\
+\ 26 \\
\hline
651
\end{array}
$$
6 + 0 is 6 (hundreds). Put 6 in the hundreds column.

The answer is six hundred fifty-one.

What happens if the tens column adds up to 10 tens?

Example   Add 625 and 77.

```
  1
 625
+ 77
   2
```
$5 + 7$ is 12 (2 ones and 1 ten). Write "2" in the ones column, and carry 1 (ten).

```
 1 1
 625
+ 77
  02
```
$1 + 2 + 7$ is 10 tens (0 tens and 1 hundred). Write 0 in the tens column and carry the 1 (hundred).

```
 1 1
 625
+ 77
 702
```
Add the 1 (hundred) to the 6 (hundreds).

The answer is seven hundred two.

What happens if the hundreds column adds up to 10 hundreds?

Example   Add 976 and 113.

```
  976
+ 113
1,089
```
9 (hundreds) + 1 (hundred) is 10 hundreds (0 hundreds and 1 thousand). Write "0" in the hundreds column and a "1" next to it. You don't have to carry.

You can carry all the way across a number.

Example   Add 498 + 715.

```
 1 1
  498
+ 715
1,213
```
$8 + 5$ is 13. Write "3" in the ones column and carry 1 (ten). $1 + 9 + 1$ (tens) is 11 tens (1 hundred and 1 ten). Write "1" in the tens column and carry the 1 (hundred).
Then $1 + 4 + 7$ (hundreds) is 12 hundreds (2 hundreds and 1 thousand).
The answer is one thousand, two hundred thirteen.

You can carry more than 1.

Example   Add 619 + 895 + 86. (Whew!)

To make it easier to add each column, add the top number to the next number, and then add each number below in order.

       2
     619   9 + 5 + 6 is 14 + 6 is 20 (0 ones and 2 tens). ⟶   9
     895   + 5
+   86   14
       0   Write "0" and carry 2 (tens).   + 6
                                                                        20

      2 2
     619   2 + 1 + 9 + 8 is 3 + 9 + 8 is 12 + 8 is 20   2
     895   (0 tens and 2 hundreds). ⟶   + 1
+   86   3
      00   Write "0" in the tens column, and carry   + 9
                2 (hundreds).   12
                                                                        + 8
                                                                        20

      2 2
     619   2 + 6 + 8 is 8 + 8 is 16 (6 hundreds and   2
     895   1 thousand). ⟶   + 6
+   86   8
   1,600   Write "6" in the hundreds column, and "1" in the   + 8
                thousands column. The answer is one thousand,   16
                six hundred.

Does it matter if a smaller number is on top or in the middle? No. Just be sure to line up the numbers on the right-hand side.

Example   Add 485 + 37 + 654.

      1 1
     485   5 + 7 + 4 is 12 + 4 is 16 (6 ones and 1 ten). Carry 1 to
      37   the tens column.
+   654   1 + 8 + 3 + 5 is 9 + 3 + 5 is 12 + 5 is 17 (7 tens and 1
   1,176   hundred). Carry 1 to the hundreds column.
                1 + 4 + 0 + 6 is 1 + 4 + 6 is 5 + 6 is 11 (1 hundred and
                1 thousand).
                The answer is one thousand, one hundred seventy-six.

Now try the exercises on the next page.

Exercises

1.	402 34 + 12	**2.**	729 10 + 31	**3.**	825 13 + 61	**4.**	335 40 + 75

5.	22 170 + 35	**6.**	84 23 + 73	**7.**	14 + 48	**8.**	3 + 27

9.	5 16 +371	**10.**	389 403 + 21

11. 80 + 20 + 30

12. 4,034 + 200 + 567 + 3

13. 64 + 830 + 1,277

14. Experts studying a family of killer whales guessed their weights as follows: mother, 13,500 pounds; father, 12,800 pounds; two babies, 1,250 pounds and 975 pounds. If the experts are right, what is the total weight of the whale family?

15. Bill's neighborhood group wanted to find out how much money they were paying for property taxes. Bill's taxes were $960. The corner variety store paid $2,635, and the houses across the street paid $896 and $784. The houses on each side of Bill's were taxed at $1,148 and $1,009. The tax on a vacant lot was only $296. How much money was paid in taxes altogether?

16. A Scout leader was about to take his group across a rope bridge over a stream. A sign warned them that the bridge could hold only 600 pounds safely. The leader weighed 170 pounds, and the boys weighed 106, 89, 116, and 78 pounds. One of the boys brought along his dog, which weighed 47 pounds. Was the group under the weight limit?

Answers

1. 448 **2.** 770 **3.** 899 **4.** 450 **5.** 227
6. 180—18 tens is the same as 1 hundred and 8 tens, so the "1" carries over to the hundreds column.
7. 62 **8.** 30 **9.** 392 **10.** 813
11. 130 **12.** 4,804 **13.** 2,171
14. 28,525 pounds **15.** $7,728 **16.** No, they were 6 pounds over the limit.

Here's your chance to do more practice before the test coming up.

Review Exercises, Adding

1. 8
   +5

2. 7
   +19

3. 2
   +28

4. 30
   +46

5. 19
   +87

6. 314
   + 71

7. 526
   + 70

8. 524
   +425

9. 402
   +688

10. 930
   +481

11. 8
    5
   +3

12. 4
    7
   + 10

13. 34
    8
   +41

14. 67
    89
   + 2

15. 92
    93
   +94

16. 103
    24
   + 41

17. 19
    10
   +288

18. 270
    608
   + 31

19. 602
    603
   +605

20. 482
    417
   +181

21. 300 + 100 + 50 + 50

22. 722 + 64 + 537 + 1,081

23. 8,271 + 467 + 6,385 + 124

24. 3,910 + 7,368 + 422 + 6,807

25. A traveling salesman put the following miles on his car: 51,000 the first year, 92,321 the next year, 14,075 the third year, and 43,156 the last year. How many miles did he travel altogether?

26. A ferryboat crossing Lake Erie from Ontario to Ohio carried 5 cars and a gravel truck. The cars weighed 3,250, 4,820, 2,875, 3,481, and 5,167 pounds. The gravel truck weighed 12,982 pounds. If the passengers and crew weighed 3,429 pounds, how much weight was the boat carrying?

Answers

25. 200,552 miles **26.** 36,004 pounds
21. 500 **22.** 2,404 **23.** 15,247 **24.** 18,507
16. 168 **17.** 317 **18.** 909 **19.** 1,810 **20.** 1,080
11. 16 **12.** 21 **13.** 83 **14.** 158 **15.** 279
6. 385 **7.** 596 **8.** 949 **9.** 1,090 **10.** 1,411
1. 13 **2.** 26 **3.** 30 **4.** 76 **5.** 106

Review Exercises, Naming Numbers

Write the following numbers in words.

1. 1,206,000 _____

2. 3,214,000 _____

3. 1,305,219 _____

_____

4. 500,400 _____

5. 3,300 _____

6. 1,230,600 _____

_____

7. 520,000,000,000 _____

8. 2,193,000,000 _____

Put the following words in number form.

9. Six billion, three hundred twenty-four million, seven hundred five thousand, one hundred ninety-eight _____

10. Three hundred sixty thousand, four hundred ninety-one _____

11. Ninety-three million _____

12. Two trillion, forty-six _____

13. Seven hundred six thousand, seven hundred six _____

14. Fifty thousand, nine hundred one _____

15. Two hundred four million _____

Answers

1. One million, two hundred six thousand   2. Three million, two hundred fourteen thousand   3. One million, three hundred five thousand, two hundred nineteen   4. Five hundred thousand, four hundred   5. Three thousand three hundred   6. One million, two hundred thirty thousand, six hundred   7. Five hundred twenty billion   8. Two billion, one hundred ninety-three million
9. 6,324,705,198   10. 360,491   11. 93,000,000
12. 2,000,000,000,046   13. 706,706   14. 50,901   15. 204,000,000

Unit 1 Test

Use a separate piece of paper to work out each problem.
Write your answers here. ⟶

Add:

1.	89	2.	47	3.	976	4.	28
	72		117		82		865
	+14		+903		+400		+354

5. 7,141 + 6,454 + 2,884 + 1,462

6. 237 + 942   **7.** 612 + 80 + 178

8. 509 + 400 + 67   **9.** 77 + 72 + 71

10. 1,042 + 1,608   **11.** 267 + 2,267

12. Allan Brown got A's on all of his final exams. He figured that he spent 186 minutes studying for his science exam, 147 minutes studying for his math exam, 295 minutes for English, and 9 minutes for his automotive course. Find the total time he spent studying.

13. A group of farmers decided to have a swamp enlarged into an irrigation pond. The drag line cost $1200 to rent and $568 for the operator's wages. Hauling the dirt away cost $1,364, gas and oil cost $104, and a building permit was $49. A fence around the hole cost $285. How much did the pond cost altogether?

Write the following words as numbers.

14. One million, two hundred sixty thousand

15. Three million, four hundred twenty-five thousand, three hundred forty-eight

16. Eight hundred thirty-two million, six hundred five thousand

17. Nine million, one hundred thirty-one thousand, one hundred fifty-four

Write the following numbers as words.

18. 250,000   **19.** 5,146,298   **20.** 198,612

Your Answers

1. _____
2. _____
3. _____
4. _____
5. _____
6. _____
7. _____
8. _____
9. _____
10. _____
11. _____
12. _____
13. _____
14. _____
15. _____
16. _____
17. _____
18. _____
_____
_____
19. _____
_____
_____
20. _____
_____
_____

2 | Subtracting Whole Numbers

Aims you toward:

- Subtracting numbers and getting the right answer.
- Knowing how to check your answer to a subtraction problem.

In this unit you will work exercises like these:

- Subtract: 363 − 97.
- Subtract and check your answer: 1,200 − 248.

Subtracting Numbers

It makes sense to talk about subtraction right after we've finished with addition. Why? Because one way to solve subtraction problems is to change them to addition problems.

A subtraction problem like this:
$$\begin{array}{r} 6 \\ -2 \\ \hline ? \end{array}$$

can be changed to an addition problem like this:
$$\begin{array}{r} 2 \\ +? \\ \hline 6 \end{array}.$$

Here we are asking, "What number can be added to 2 to get 6?"

We know that
$$\begin{array}{r} 2 \\ +4 \\ \hline 6 \end{array}.$$
This tells us that
$$\begin{array}{r} 6 \\ -2 \\ \hline 4 \end{array}.$$

(Later in this unit, you'll see how adding can be used to check your answer to a subtraction problem.)

Let's do some subtraction problems to make sure you have the right idea.

Examples

$$\begin{array}{r} 8 \\ -3 \\ \hline \end{array}$$
We ask
$$\begin{array}{r} 3 \\ +? \\ \hline 8 \end{array}.$$
We know that
$$\begin{array}{r} 3 \\ +5 \\ \hline 8 \end{array},$$
so
$$\begin{array}{r} 8 \\ -3 \\ \hline 5 \end{array}.$$

$$\begin{array}{r} 9 \\ -6 \\ \hline 3 \end{array}$$
because
$$\begin{array}{r} 6 \\ +3 \\ \hline 9 \end{array}$$
$$\begin{array}{r} 7 \\ -2 \\ \hline 5 \end{array}$$
because
$$\begin{array}{r} 2 \\ +5 \\ \hline 7 \end{array}$$

$$\begin{array}{r} 15 \\ -8 \\ \hline 7 \end{array}$$
because
$$\begin{array}{r} 8 \\ +7 \\ \hline 15 \end{array}$$
$$\begin{array}{r} 13 \\ -4 \\ \hline 9 \end{array}$$
because
$$\begin{array}{r} 4 \\ +9 \\ \hline 13 \end{array}$$

What happens when you want to subtract larger numbers?

Example Subtract 53 − 12. When you see a − sign you need to SUBTRACT.

(minuend)
(subtrahend)
$$\begin{array}{r} 53 \\ -12 \\ \hline \end{array}$$
Line up the numbers on the right-hand side. Always put the "take away" number (subtrahend) on the bottom.

$$\begin{array}{r} 53 \\ -12 \\ \hline 1 \end{array}$$
Subtract the bottom number from the top number in the ones column.

(difference)
$$\begin{array}{r} 53 \\ -12 \\ \hline 41 \end{array}$$
Go to the tens column and subtract. Remember to keep the columns lined up! The answer is 41.

Exercises

1. $\begin{array}{r} 5 \\ -2 \end{array}$	2. $\begin{array}{r} 7 \\ -4 \end{array}$	3. $\begin{array}{r} 9 \\ -5 \end{array}$	4. $\begin{array}{r} 16 \\ -\ 9 \end{array}$	5. $\begin{array}{r} 13 \\ -\ 6 \end{array}$
6. $\begin{array}{r} 12 \\ -\ 5 \end{array}$	7. $\begin{array}{r} 3 \\ -1 \end{array}$	8. $\begin{array}{r} 6 \\ -2 \end{array}$	9. $\begin{array}{r} 8 \\ -4 \end{array}$	10. $\begin{array}{r} 5 \\ -3 \end{array}$
11. $\begin{array}{r} 14 \\ -\ 8 \end{array}$	12. $\begin{array}{r} 17 \\ -\ 8 \end{array}$	13. $\begin{array}{r} 82 \\ -41 \end{array}$	14. $\begin{array}{r} 98 \\ -73 \end{array}$	15. $\begin{array}{r} 65 \\ -24 \end{array}$
16. $\begin{array}{r} 49 \\ -38 \end{array}$	17. $\begin{array}{r} 73 \\ -63 \end{array}$	18. $\begin{array}{r} 19 \\ -\ 9 \end{array}$	19. $\begin{array}{r} 57 \\ -54 \end{array}$	20. $\begin{array}{r} 94 \\ -21 \end{array}$
21. $\begin{array}{r} 484 \\ -\ 32 \end{array}$	22. $\begin{array}{r} 766 \\ -\ 42 \end{array}$	23. $\begin{array}{r} 835 \\ -\ 35 \end{array}$	24. $\begin{array}{r} 690 \\ -260 \end{array}$	25. $\begin{array}{r} 395 \\ -163 \end{array}$

Answers

1. 3	2. 3	3. 4	4. 7	5. 7
6. 7	7. 2	8. 4	9. 4	10. 2
11. 6	12. 9	13. 41	14. 25	15. 41
16. 11	17. 10	18. 10	19. 3	20. 73
21. 452	22. 724	23. 800	24. 430	25. 232

By now you've learned quite a bit about
subtraction, but stay tuned, because
you're going to learn more.

Suppose you want to buy a stereo system for $363. You talk the sales clerk into giving
you a $45 discount. Now how much will you have to pay?

$$\begin{array}{r} 363 \\ -\ 45 \end{array}$$

Start as usual with the ones column—
but you've got a problem—
The 5 is too big to take away from the 3!

But there are 6 tens just sitting there.

$$\begin{array}{r} \overset{5}{3}\cancel{6}\overset{1}{3} \\ -\ 45 \end{array}$$

Borrow one of the tens.
Cross out the 6 and put a 5 above to show you now
have only 5 tens.
How many ones is the borrowed 10? 10 ones.

Add the 10 ones to the 3 ones.
You now have 13 ones.
Put a 1 by the 3 to show you now have 13 ones.

$$\begin{array}{r} \overset{5}{3}\cancel{6}{}^1 3 \\ -\ 4\ 5 \\ \hline 8 \end{array}$$

Now you can subtract!
Begin with the ones column and move left.
13 − 5 is 8.

$$\begin{array}{r} \overset{5}{3}\cancel{6}{}^1 3 \\ -\ 4\ 5 \\ \hline 1\ 8 \end{array}$$

5 − 4 is 1.

$$\begin{array}{r} \overset{5}{3}\cancel{6}{}^1 3 \\ -\ 4\ 5 \\ \hline 31\ 8 \end{array}$$

3 − 0 is 3.
That's all there is to it.
The stereo system will cost you $318.

Let's do some more exercises.

Exercises

1. $\overset{5}{\cancel{6}}{}^{1}4$
    $-\ 3\ 9$

2. 33
    -19

3. 78
    -49

4. 55
    -26

5. 92
    -67

6. 846
    -328

7. 775
    -738

8. 693
    -467

9. 384
    $-\ \ 26$

10. 547
    -339

11. $56 - 39$

$\overset{4}{\cancel{5}}{}^{1}6$
$-\ 3\ 9$

12. $78 - 49$

13. $54 - 27$

14. $61 - 33$

15. $93 - 38$

16. $28 - 19$

17. $36 - 18$

18. $72 - 56$

19. $837 - 419$

$\overset{2}{8}\cancel{3}{}^{1}7$
$-41\ 9$

20. $144 - 126$

21. $455 - 316$

22. $396 - 178$

23. Marjorie is driving to a city 874 miles away. She has already driven 366 miles. How much farther does she have to drive?

24. A city police force numbered 612. During a flu epidemic in March, only 487 officers reported for duty. How many were off sick?

Answers

23. 508 miles **24.** 125 police officers

21. 139 **22.** 218

16. 9 **17.** 18 **18.** 16 **19.** 418 **20.** 18

11. 17 **12.** 29 **13.** 27 **14.** 28 **15.** 55

6. 518 **7.** 37 **8.** 226 **9.** 358 **10.** 208

1. 25 **2.** 14 **3.** 29 **4.** 29 **5.** 25

Subtracting Large Numbers

Let's take a look again at that $363 stereo system you want to buy. Suppose now that you're a real fast talker and the sales clerk agrees to give you a $97 discount. How much will you have to pay?

$$
\begin{array}{r}
363 \\
-\ 97 \\
\end{array}
$$

The 7 is bigger than the 3, so you know you have to borrow.

$$
\begin{array}{r}
3\ \overset{5}{\cancel{6}}{}^{1}3 \\
-\ \ 9\ 7 \\
\end{array}
$$

Borrow one of the tens. (This leaves you 5 tens.)
10 is 10 ones.
Add the 10 ones to the 3 ones.

$$
\begin{array}{r}
3\ \overset{5}{\cancel{6}}{}^{1}3 \\
-\ \ 9\ 7 \\
\hline
?\ 6 \\
\end{array}
$$

Now subtract. 13 − 7 is 6.
The 9 is bigger than the 5.
Any ideas about what to do?

Take a look at the 3 hundreds.

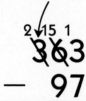

Borrow one of the hundreds.
Cross out the 3 and put a 2 above to show you now have only 2 hundreds.
How many tens is the borrowed 100? 10 tens.

Add the 10 tens to the 5 tens.
You now have 15 tens.
Put a 1 by the 5 to show you now have 15 tens.

Now you can finish subtracting.

$$
\begin{array}{r}
\overset{2}{\cancel{3}}\ \overset{15}{\cancel{6}}{}^{1}3 \\
-\ \ 9\ 7 \\
\hline
6\ 6 \\
\end{array}
$$

15 − 9 is 6.

$$
\begin{array}{r}
\overset{2}{\cancel{3}}\ \overset{15}{\cancel{6}}{}^{1}3 \\
-\ \ 9\ 7 \\
\hline
2\ 6\ 6 \\
\end{array}
$$

2 − 0 is 2.

Exercises are next.

Exercises

1. 332 − 43	**2.** 165 − 97	**3.** 516 − 37	**4.** 118 − 37	**5.** 987 − 88
6. 142 − 53	**7.** 214 − 55	**8.** 172 − 86	**9.** 912 − 37	**10.** 420 − 66

11. 524 − 35   **12.** 911 − 72   **13.** 463 − 85   **14.** 371 − 92

15. A car had a list price of $8,722. If a discount of $686 was given, how much was paid for the car?

16. Last year a delivery van used 1,500 gallons of gas. A service station mechanic told the owner to keep his engine tuned in order to get better mileage. He took the mechanic's advice, and this year he used only 985 gallons of gas for the same amount of work. How many gallons of gas did he save?

Answers

16. 515 gallons

11. 489	**12.** 839	**13.** 378	**14.** 279	**15.** $8,036
6. 89	**7.** 159	**8.** 86	**9.** 875	**10.** 354
1. 289	**2.** 68	**3.** 479	**4.** 81	**5.** 899

Sometimes you have to take a *double* step when you subtract.

Example   Subtract 24 from 301.

```
  301        The 4 is bigger than the 1, so you know you have to borrow.
 − 24        But there's a zero in the tens place.
             You need some tens before you can borrow.
```

```
   2
  3ˌ¹0 1     So you have to go over to the next number and borrow
 −   2 4     some tens. 100 is 10 tens.
             10 tens plus 0 tens is 10 tens.
```

```
   2
  ꓱ 0 1      Now you have tens to borrow from.
 −   2 4     You have 9 tens left.
```

```
   2  9
  ꓱ ꓳ¹1      Now you can subtract.
 −   2 4
   2 7 7
```

Examples

```
   7 9            3 9 9            2 9 9            4 9            5 10 9
  ꓷ ꓳ¹0         4 ꓳ ꓳ¹1         ꓱ ꓳ ꓳ¹0         5 ꓳ¹7          6 ꓲ ꓳ¹4
 −2 2 9         −  3 5 2         −  8 1 7         −2 9 8         −  8 7 7
  5 7 1          3,6 4 9          2,1 8 3          2 0 9          5,2 2 7
```

Let's try some bigger numbers. Don't panic!

Example
$$\begin{array}{r} 25{,}504 \\ -\ 6{,}378 \end{array}$$

Begin with the ones column.
8 is bigger than 4, so you must borrow from the tens.
But there are no tens.

$$\begin{array}{r} 2\ 5{,}\overset{4}{\cancel{5}}\overset{1}{0}\ 4 \\ -\ 6{,}3\ 7\ 8 \end{array}$$

You get some tens by borrowing one of the hundreds.
100 is 10 tens.
10 tens plus 0 tens is 10 tens.

$$\begin{array}{r} 2\ 5{,}\overset{4}{\cancel{5}}\overset{9}{\cancel{0}}\overset{1}{4} \\ -\ 6{,}3\ 7\ 8 \end{array}$$

Now you have tens to borrow from.
10 is 10 ones.
10 ones plus 4 ones is 14 ones.

$$\begin{array}{r} 2\ 5{,}\overset{4}{\cancel{5}}\overset{9}{\cancel{0}}\overset{1}{4} \\ -\ 6{,}3\ 7\ \underline{8} \\ 6 \end{array}$$

Now subtract.
14 − 8 is 6.

$$\begin{array}{r} 2\ 5{,}\overset{4}{\cancel{5}}\overset{9}{\cancel{0}}\overset{1}{4} \\ -\ 6{,}3\ 7\ 8 \\ 2\ 6 \end{array}$$

9 − 7 is 2.

$$\begin{array}{r} 2\ 5{,}\overset{4}{\cancel{5}}\overset{9}{\cancel{0}}\overset{1}{4} \\ -\ 6{,}3\ 7\ 8 \\ 1\ 2\ 6 \end{array}$$

4 − 3 is 1.

$$\begin{array}{r} \overset{1}{\cancel{2}}{}^{1}5{,}\overset{4}{\cancel{5}}\overset{9}{\cancel{0}}\overset{1}{4} \\ -\ 6{,}3\ 7\ 8 \\ 1\ 2\ 6 \end{array}$$

In the thousands columns, you see that 6 is bigger than 5. Borrow one of the ten thousands.
10,000 is 10 thousands.
10 thousands plus 5 thousands is 15 thousands.

$$\begin{array}{r} \overset{1}{\cancel{2}}{}^{1}5{,}\overset{4}{\cancel{5}}\overset{9}{\cancel{0}}\overset{1}{4} \\ -\ 6{,}3\ 7\ 8 \\ 9{,}1\ 2\ 6 \end{array}$$

Finish subtracting.
15 − 6 is 9.

$$\begin{array}{r} \overset{1}{\cancel{2}}{}^{1}5{,}\overset{4}{\cancel{5}}\overset{9}{\cancel{0}}\overset{1}{4} \\ -\ 6{,}3\ 7\ 8 \\ 1\ 9{,}1\ 2\ 6 \end{array}$$

1 − 0 is 1.

Exercises are on the next page. ▶

Exercises

1. 472 − 399	**2.** 78,114 − 3,550	**3.** 22,397 −13,499	**4.** 909 − 10
5. 6,625 − 837	**6.** 700 − 296	**7.** 3,400,214 −1,621,986	**8.** 8,511,343 −4,100,679
9. 89,603 −77,999	**10.** 59,203,759 − 6,156,699	**11.** 609,483,652 −437,598,368	**12.** 99,372,174 − 98,386

Line the numbers up, with the first number (minuend) on the top.

13. 983 − 899   **14.** 540 − 368   **15.** 78,945 − 69,066

16. 33,782 − 21,893   **17.** 74,839 − 66,944   **18.** 983 − 495

19. Eric's income is $9,814 per year. He pays $1,028 in taxes. What is his take-home pay?

20. Several years ago a small town had a population of 8,695 people. Since then, 3 new factories and a hospital were built, and the population jumped to 12,000. What was the population increase?

Answers

20. 3,305 people

19. $8,786

18. 488

17. 7,895

16. 11,889   **15.** 9,879   **14.** 172   **13.** 84

12. 99,273,788   **11.** 171,885,284   **10.** 53,047,060   **9.** 11,604

8. 4,410,664   **7.** 1,778,228   **6.** 404   **5.** 5,788

4. 899   **3.** 8,898   **2.** 74,564   **1.** 73

Checking Answers

Congratulations! You've now learned the basic skills of subtraction. So far you've been checking your answers by comparing them with the answers at the bottom of the page. There's another way to check your answers.

When you have a problem like
$$\begin{array}{r} 78 \\ -\,34 \\ \hline \end{array}$$
you can now easily get the answer.

$$\begin{array}{r} 78 \\ -\,34 \\ \hline 44 \end{array}$$ ← Cover this number for a minute.

$$\begin{array}{r} 34 \\ +\,44 \\ \hline 78 \end{array}$$ Add your answer to 34.

Does 78 look familiar?

$$\begin{array}{r} 78 \\ -\,34 \\ \hline 44 \end{array} \qquad \begin{array}{r} 34 \\ +\,44 \\ \hline 78 \end{array}$$

That's not magic; that's arithmetic!

Checking a Subtraction Problem	Take your answer (difference) and add it to the second number (subtrahend) in the problem. It should equal the first number (minuend) in the problem. If it doesn't, your answer is wrong, but don't give up. Just start the subtraction problem over.

Let's run through a few examples of checking subtraction problems by adding.

Examples

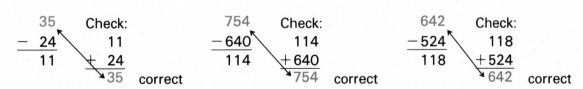

$$\begin{array}{r} 35 \\ -\,24 \\ \hline 11 \end{array}$$ Check:
$$\begin{array}{r} 11 \\ +\,24 \\ \hline 35 \end{array}$$ correct

$$\begin{array}{r} 754 \\ -\,640 \\ \hline 114 \end{array}$$ Check:
$$\begin{array}{r} 114 \\ +\,640 \\ \hline 754 \end{array}$$ correct

$$\begin{array}{r} 642 \\ -\,524 \\ \hline 118 \end{array}$$ Check:
$$\begin{array}{r} 118 \\ +\,524 \\ \hline 642 \end{array}$$ correct

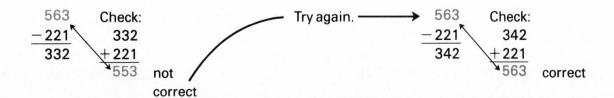

$$\begin{array}{r} 563 \\ -\,221 \\ \hline 332 \end{array}$$ Check:
$$\begin{array}{r} 332 \\ +\,221 \\ \hline 553 \end{array}$$ not correct

Try again. ⟶

$$\begin{array}{r} 563 \\ -\,221 \\ \hline 342 \end{array}$$ Check:
$$\begin{array}{r} 342 \\ +\,221 \\ \hline 563 \end{array}$$ correct

Exercises

Check each of the following problems by adding. Some are correct; some are not correct.

1.	56	2.	948	3.	6,746	4.	9,962
	-37		$-\ 33$		$-3,082$		$-9,602$
	19		916		3,664		360

5.	7,493	6.	632,408	7.	5,432	8.	7,952
	$-6,394$		$-\ 26,743$		$-5,178$		$-1,503$
	1,109		505,765		254		6,449

Work each subtraction problem. Check each answer by adding.

9. $347 - 298$ **10.** $1,362 - 375$ **11.** $260 - 172$

12. $903 - 56$ **13.** $871 - 85$ **14.** $1,200 - 248$

15. $20,510 - 7,438$ **16.** $50,982 - 23,661$

17. Tracey Everitt bought some topsoil for her garden last year. The delivery truck held 8,152 pounds of topsoil when it arrived, and it held 5,763 pounds of topsoil after Tracey's shipment was unloaded. How many pounds of topsoil did Tracey buy?

18. Twenty years ago Bernie and Nancy paid $8,955 for their house. Today a real estate firm says that it is worth $41,250. How much did the property gain in value?

Answers

17. 2,389 pounds   **18.** $32,295

12. 847 **13.** 786 **14.** 952 **15.** 13,072 **16.** 27,321

6. Not correct **7.** Correct **8.** Correct **9.** 49 **10.** 987 **11.** 88

1. Correct **2.** Not correct **3.** Correct **4.** Correct **5.** Not correct

Review Exercises

Work each subtraction problem. Check each answer by adding.

1. 847
  −235

2. 9,427
  −8,304

3. 189
  − 52

4. 3,974
  −1,652

5. 1,263
  − 375

6. 924
  − 38

7. 481
  − 57

8. 2,613
  − 893

9. 836
  −798

10. 4,781
  − 869

11. 1,824
  − 957

12. 4,000
  −2,181

13. 698 − 59

14. 250 − 19

15. 100 − 70

16. 3,204
  − 597

17. 25,712
  − 6,333

18. 5,279
  − 690

19. 8,010
  −5,478

20. Mr. Dobos hauled 9,272 pounds of tobacco from his plantation to the auction barns. He sold 6,428 pounds, and he had to bring home the rest for next week's sale. How much tobacco did he carry home?

21. A publishing company published a dictionary with 16,637 words in it. Later it published a pocket-sized version with 9,638 words. How many words were left out?

22. Heather earned $1,100 for her first summer job. Out of this, she had to pay her mother $138 for last year's long-distance telephone calls. How much did she have left?

Answers

20. 2,844 pounds 21. 6,999 words 22. $962
16. 2,607 17. 19,379 18. 4,589 19. 2,532
11. 867 12. 1,819 13. 639 14. 231 15. 30
6. 886 7. 424 8. 1,720 9. 38 10. 3,912
1. 612 2. 1,123 3. 137 4. 2,322 5. 888

Unit 2 Test

Use a separate piece of paper to work out each problem.
Check each answer by adding.

Subtract:

1.   2,864
  $-$ 743

2.   4,783
  $-$ 251

3.   82
  -26

4.   5,192
  $-$ 304

5.   47,821
  $-$ 8,939

6.   4,700
  $-$ 89

7.   28,004
  $-$ 3,608

8.   50,982
  $-23,661$

9.   30,112
  $-$ 9,873

10.   83,144
  $-$ 2,497

11. $325 - 14$

12. $95 - 35$

13. $72 - 68$

14. $574 - 231$

15. $5,421 - 85$

16. $4,905 - 299$

17. $18,406 - 9,520$

18. $45,011 - 21,539$

19. Last year, Josie's travel agency showed a net profit
(which is all the money left over *after* she takes out her
expenses) of $61,497.86. Her gross profit (all the
money she made *before* any expenses were taken out)
was $101,465.19. How much were her expenses?

20. One week, Bonnie read 1,243 pages of a set of books
on home nursing. She planned to finish reading the set
during the following week, when she would be on
vacation. If the set of books had 2,960 pages altogether,
how many pages were left to read?

Your Answers

1. _____

2. _____

3. _____

4. _____

5. _____

6. _____

7. _____

8. _____

9. _____

10. _____

11. _____

12. _____

13. _____

14. _____

15. _____

16. _____

17. _____

18. _____

19. _____

20. _____

3 | Multiplying Whole Numbers

Aims you toward:

- Knowing the multiplication tables.
- Multiplying numbers and getting the right answer.

In this unit you will work exercises like these:

- Multiply 7×8.
- Multiply 423×2.
- Multiply 362×214.

Getting Ready to Multiply

After working so many problems in addition and subtraction in the first two units, you're probably ready to learn a few shortcuts. If that's true, this is the unit for you, because . . .

MULTIPLICATION is a SHORTCUT for ADDITION.

Suppose you take 4 coffee breaks during the day. Each cup costs 50¢. How much do you spend each day for coffee?

If you were to solve this problem the way we talked about in Unit 1, you would probably solve the problem like this:

```
    50¢
    50
    50
 +  50
    200¢ altogether
```

BUT, we also have a shortcut approach:

$$
\begin{array}{r}
50¢ \quad \text{(factor)} \\
\times \quad 4 \quad \text{(factor)} \\
\hline
200¢ \quad \text{(product)}
\end{array}
$$

Factors are *any* numbers that will give the answer 200 when you multiply them.

$$50 \times 4 = 200$$

> Stop for a moment. Look at the = sign used above. It is read "is equal to." So when you see $50 \times 4 = 200$, it is the same as writing 50 times 4 is equal to 200, or 50×4 is 200. When you see an = sign, you know that whatever is on the left of the sign is equal to what is on the right of the sign.

How do we know that $50 \times 4 = 200$? Before we can answer that question, there's something you must do. ⟶

BEFORE YOU CAN GO ANY FURTHER

IN MATHEMATICS

YOU MUST KNOW

THE MULTIPLICATION TABLES!

LEARN THEM ANY WAY YOU CAN!

They're on the next page . . .
They simply *must* be memorized;
there is no way out of it.

Multiplication Tables

$1 \times 1 = 1$	$2 \times 1 = 2$	$3 \times 1 = 3$
$1 \times 2 = 2$	$2 \times 2 = 4$	$3 \times 2 = 6$
$1 \times 3 = 3$	$2 \times 3 = 6$	$3 \times 3 = 9$
$1 \times 4 = 4$	$2 \times 4 = 8$	$3 \times 4 = 12$
$1 \times 5 = 5$	$2 \times 5 = 10$	$3 \times 5 = 15$
$1 \times 6 = 6$	$2 \times 6 = 12$	$3 \times 6 = 18$
$1 \times 7 = 7$	$2 \times 7 = 14$	$3 \times 7 = 21$
$1 \times 8 = 8$	$2 \times 8 = 16$	$3 \times 8 = 24$
$1 \times 9 = 9$	$2 \times 9 = 18$	$3 \times 9 = 27$
$4 \times 1 = 4$	$5 \times 1 = 5$	$6 \times 1 = 6$
$4 \times 2 = 8$	$5 \times 2 = 10$	$6 \times 2 = 12$
$4 \times 3 = 12$	$5 \times 3 = 15$	$6 \times 3 = 18$
$4 \times 4 = 16$	$5 \times 4 = 20$	$6 \times 4 = 24$
$4 \times 5 = 20$	$5 \times 5 = 25$	$6 \times 5 = 30$
$4 \times 6 = 24$	$5 \times 6 = 30$	$6 \times 6 = 36$
$4 \times 7 = 28$	$5 \times 7 = 35$	$6 \times 7 = 42$
$4 \times 8 = 32$	$5 \times 8 = 40$	$6 \times 8 = 48$
$4 \times 9 = 36$	$5 \times 9 = 45$	$6 \times 9 = 54$
$7 \times 1 = 7$	$8 \times 1 = 8$	$9 \times 1 = 9$
$7 \times 2 = 14$	$8 \times 2 = 16$	$9 \times 2 = 18$
$7 \times 3 = 21$	$8 \times 3 = 24$	$9 \times 3 = 27$
$7 \times 4 = 28$	$8 \times 4 = 32$	$9 \times 4 = 36$
$7 \times 5 = 35$	$8 \times 5 = 40$	$9 \times 5 = 45$
$7 \times 6 = 42$	$8 \times 6 = 48$	$9 \times 6 = 54$
$7 \times 7 = 49$	$8 \times 7 = 56$	$9 \times 7 = 63$
$7 \times 8 = 56$	$8 \times 8 = 64$	$9 \times 8 = 72$
$7 \times 9 = 63$	$8 \times 9 = 72$	$9 \times 9 = 81$

In learning your tables, you actually learn two products for every one that you memorize. For example:

$$4 \times 8 = 32 \quad \text{and} \quad 8 \times 4 = 32.$$

Also remember that any number times zero is zero, and zero times any number is zero.

Example If you're not sure of the answers to, say,

$$6 \times 0 = ? \quad \text{or} \quad 28 \times 0 = ?$$

think of

$$6 \times 1 = 6 \quad \text{and} \quad 28 \times 1 = 28.$$

Since you can't have the same answers when you multiply by 0 and by 1, 6×0 has to be 0, and 28×0 is always 0.

Now that you've memorized your multiplication tables, you can relax for a few minutes.

BUT NOT TOO LONG! ! !

Try the problems on the next page to see if you can remember the tables.

Exercises

Multiply. Try to do these without looking at the multiplication tables.

1. $3 \times 9 =$ _____ **2.** $4 \times 0 =$ _____ **3.** $7 \times 4 =$ _____ **4.** $5 \times 7 =$ _____

5. $9 \times 9 =$ _____ **6.** $8 \times 4 =$ _____ **7.** $3 \times 6 =$ _____ **8.** $5 \times 8 =$ _____

9. $9 \times 4 =$ _____ **10.** $4 \times 4 =$ _____ **11.** $8 \times 3 =$ _____ **12.** $3 \times 5 =$ _____

13. $5 \times 4 =$ _____ **14.** $8 \times 8 =$ _____ **15.** $8 \times 5 =$ _____ **16.** $4 \times 8 =$ _____

17. $3 \times 8 =$ _____ **18.** $6 \times 4 =$ _____ **19.** $6 \times 7 =$ _____ **20.** $4 \times 3 =$ _____

21. $4 \times 5 =$ _____ **22.** $1 \times 6 =$ _____ **23.** $3 \times 7 =$ _____ **24.** $5 \times 6 =$ _____

25. $6 \times 8 =$ _____ **26.** $3 \times 4 =$ _____ **27.** $6 \times 9 =$ _____ **28.** $7 \times 7 =$ _____

29. $7 \times 6 =$ _____ **30.** $4 \times 9 =$ _____ **31.** $9 \times 3 =$ _____ **32.** $7 \times 5 =$ _____

33. $5 \times 3 =$ _____ **34.** $5 \times 5 =$ _____ **35.** $8 \times 9 =$ _____ **36.** $9 \times 6 =$ _____

37. $6 \times 5 =$ _____ **38.** $4 \times 7 =$ _____ **39.** $6 \times 6 =$ _____ **40.** $4 \times 6 =$ _____

Answers

		40. 24	
36. 54	**35.** 72	**34.** 25	**39.** 36	**38.** 28	**37.** 30
30. 36	**29.** 42	**28.** 49	**33.** 15	**32.** 35	**31.** 27
24. 30	**23.** 21	**22.** 6	**27.** 54	**26.** 12	**25.** 48
18. 24	**17.** 24	**16.** 32	**21.** 20	**20.** 12	**19.** 42
12. 15	**11.** 24	**10.** 16	**15.** 40	**14.** 64	**13.** 20
6. 32	**5.** 81	**4.** 35	**9.** 36	**8.** 40	**7.** 18
			3. 28	**2.** 0	**1.** 27

Multiplying Numbers

If you are *sure* you have the multiplication tables memorized, you are ready to go on.

Example Suppose you are working for your uncle, a florist. A classmate who works for a catering service calls. She is making arrangements for a benefit banquet, and she needs to order 4 dozen roses. If the roses are $12 a dozen, how much will they cost? Multiply to find out.

   12   Start in the ones column.
× 4 $4 \times 2 = 8$
    8   Put the 8 in the ones column.

   12   Next go to the tens column.
× 4 $4 \times 1 = 4$
  48   Since the 1 is in the tens place, this is actually 4×10, which equals 40.
        40 is 4 tens. Put the 4 in the tens column.
        The answer is 48.

4 dozen roses at $12 a dozen will cost $48.

Another Example Multiply 287 by 3.

     2
 287   Start with the ones column.
×   3 $3 \times 7 = 21$
    1   21 is 2 tens and 1 one.
        Put the 1 in the ones column.
        Carry the 2 tens.
        Put a 2 in the tens column above the 8.

   2 2
 287   Next multiply the tens column.
×   3 $3 \times 8 = 24$
  61   Since 8 is in the tens place, this is actually 3×80.
        3×80 (8 tens) = 240
        240 is 2 hundreds and 4 tens.
        Add on the 2 tens you already have.
        This gives you 2 hundreds and 4 + 2 or 6 tens.
        Put a 6 in the tens place.
        Carry the 2 hundreds.

   2 2
 287   Next multiply the hundreds column.
×   3 $3 \times 2 = 6$
 861   3×200 (2 hundreds) = 600
        Add on the 2 hundreds you already have.
        6 hundreds + 2 hundreds = 8 hundreds.
        Put an 8 in the hundreds place.
        The answer is 861.

Confused by all this carrying and adding? Let's slow things down and take a closer look at what happens when you multiply.

Here's the problem from the last example worked out step-by-step.

```
 287
×   3
  21
```
Start with the ones column.
3 × 7 = 21
21 is 2 tens and 1 one.
Put the 1 in the ones column.
Instead of carrying the 2 tens, put a 2 in the tens place beside the 1.

```
 287
×   3
  21
 240
```
Next multiply the tens column.
3 × 8 = 24
3 × 80 (8 tens) = 240

```
 287
×   3
  21
 240
 600
```
Next multiply the hundreds column.
3 × 2 = 6
3 × 200 (2 hundreds) = 600

```
 287
×   3
  21
 240
 600
 861
```
21, 240, and 600 are called partial products.
Add the partial products.

The answer is 861.

This step-by-step method shows exactly what happens when you multiply. It should help you understand the shorter method shown on the previous page.

Compare the two methods. The shorter method takes less time once you get used to it. The examples below are worked both ways.

Examples

```
   3
  25        25        23       23        2 2
× 7   or  × 7       × 2  or  × 2        344          344
 175        35        46        6      ×   5   or  ×   5
           140                 40      1,720          20
           175                 46                    200
                                                   1,500
                                                   1,720
```

More exercises on multiplying are on the next page.

Exercises

1.
$$\begin{array}{r} {}^{1} \\ 83 \\ \times\ \ 4 \\ \hline 332 \end{array}$$

2.
$$\begin{array}{r} 41 \\ \times\ 7 \\ \hline \end{array}$$

3.
$$\begin{array}{r} 24 \\ \times\ 2 \\ \hline \end{array}$$

4.
$$\begin{array}{r} 86 \\ \times\ 6 \\ \hline \end{array}$$

5.
$$\begin{array}{r} 13 \\ \times\ 3 \\ \hline \end{array}$$

6.
$$\begin{array}{r} 39 \\ \times\ 2 \\ \hline \end{array}$$

7.
$$\begin{array}{r} 61 \\ \times\ 4 \\ \hline \end{array}$$

8.
$$\begin{array}{r} 73 \\ \times\ 2 \\ \hline \end{array}$$

9.
$$\begin{array}{r} 22 \\ \times\ 8 \\ \hline \end{array}$$

10.
$$\begin{array}{r} 53 \\ \times\ 5 \\ \hline \end{array}$$

11.
$$\begin{array}{r} 132 \\ \times\ \ 6 \\ \hline \end{array}$$

12.
$$\begin{array}{r} 446 \\ \times\ \ 3 \\ \hline \end{array}$$

13.
$$\begin{array}{r} 622 \\ \times\ \ 7 \\ \hline \end{array}$$

14.
$$\begin{array}{r} 612 \\ \times\ \ 4 \\ \hline \end{array}$$

15.
$$\begin{array}{r} 603 \\ \times\ \ 4 \\ \hline \end{array}$$

16.
$$\begin{array}{r} 521 \\ \times\ \ 4 \\ \hline \end{array}$$

17.
$$\begin{array}{r} 8013 \\ \times\ \ \ 4 \\ \hline \end{array}$$

18.
$$\begin{array}{r} 742 \\ \times\ \ 2 \\ \hline \end{array}$$

19.
$$\begin{array}{r} 317 \\ \times\ \ 7 \\ \hline \end{array}$$

20.
$$\begin{array}{r} 2043 \\ \times\ \ \ 5 \\ \hline \end{array}$$

21. Greg, a dental assistant, takes X rays of patients' teeth. On Monday, he saw 13 patients and they each required 5 X rays. How many X rays did he take?

22. Suppose you are the manager of a day camp. You decide that you will allow 13 children to attend the camp for every one counselor. If the camp has 9 counselors, how many children can attend?

Answers

21. 65 X rays   22. 117 children

1. 332   2. 287   3. 48   4. 516   5. 39
6. 78   7. 244   8. 146   9. 176   10. 265
11. 792   12. 1,338   13. 4,354   14. 2,448   15. 2,412
16. 2,084   17. 32,052   18. 1,484   19. 2,219   20. 10,215

Multiplying Large Numbers

You have learned to multiply by one-digit numbers.

You will now be multiplying by numbers that have more than one digit.

$$
\begin{array}{r}
243 \\
\times \ \ 2 \\
\hline
486
\end{array}
$$

Example   Multiply 184 by 39.

To work a problem like this, multiply 184 by each digit of 39, one digit at a time.

$$
\begin{array}{r}
7\,3 \\
184 \\
\times \ \ 39 \\
\hline
1656
\end{array}
$$

Step 1: Multiply 184 by 9.

$$
\begin{array}{r}
2\,1 \\
\cancel{7\,3} \\
184 \\
\times \ \ 39 \\
\hline
1656 \\
5520
\end{array}
$$

Step 2: Multiply 184 by 30 (3 tens).
(Write 0 in the ones place and multiply 184 by 3.)
Use new carry numbers; cross out the old ones.
$3 \times 4 = 12$ (Carry the 1.)
$3 \times 8 = 24$ $24 + 1 = 25$ (Carry the 2.)
$3 \times 1 = 1$ $3 + 2 = 5$

$$
\begin{array}{r}
2\,1 \\
\cancel{7\,3} \\
184 \\
\times \ \ 39 \\
\hline
1656 \\
5520 \\
\hline
7176
\end{array}
$$

Step 3: Now add the partial products.

The answer is 7,176.

Another Example   Multiply 68 by 14.

$$
\begin{array}{r}
3 \\
68 \\
\times \ 14 \\
\hline
272
\end{array}
$$

First multiply 68×4.

$$
\begin{array}{r}
\cancel{3} \\
68 \\
\times \ 14 \\
\hline
272 \\
680
\end{array}
$$

Next multiply 68×10 (1 ten).
Write 0 in the ones place and multiply 68×1.)

$$
\begin{array}{r}
\cancel{3} \\
68 \\
\times \ 14 \\
\hline
272 \\
680 \\
\hline
952
\end{array}
$$

Add the partial products.

The answer is 952.

Exercises

1.
```
    1 1
    2 2
   567
 ×  24
  2268
 11340
```

2.
```
   416
 ×  37
```

3.
```
   564
 ×  54
```

4.
```
  4682
 ×   83
```

5.
```
   603
 ×  87
```

6.
```
   827
    73
```

7.
```
   564
 ×  28
```

8.
```
  3316
 ×   41
```

9.
```
   182
 ×  12
```

10.
```
   347
 ×  76
```

11.
```
   198
    44
```

12.
```
  4327
 ×   53
```

13.
```
   644
 ×  55
```

14.
```
   857
 ×  92
```

15.
```
   484
 ×  73
```

16. How many hours are there in 2 years? (1 year is 365 days.)

Answers

16. 17,520 hours
1. 13,608 2. 15,392 3. 30,456 4. 388,606 5. 52,461
6. 60,371 7. 15,792 8. 135,956 9. 2,184 10. 26,372
11. 8,712 12. 229,331 13. 35,420 14. 78,844 15. 35,332

What happens when you multiply by a three-digit number? You use the same methods you just learned for multiplication by a two-digit number.

Example Multiply 594 by 278.

```
    7 3
    594          Multiply 594 by 8.
  ×  278
   4 752
```

```
    6 2
    7 3
    594          Now multiply 594 by 70.
  ×  278          (Write 0 in the ones place, and multiply 594 by 7.)
   4 752
  41 580
```

```
      1
    6 2
    7 3
    594          Now multiply 594 by 200.
  ×  278          (Write 0's in the ones and tens places, and multiply
   4 752          594 by 2.)
  41 580
 118 800
```

```
      1
    6 2
    7 3
    594          Now add the partial products.
  ×  278
   4 752
  41 580
 118 800
 165,132         The answer is 165,132.
```

Here are some more examples of multiplication by large numbers.

Examples

```
   1 1
   5 7
   2 3
    369          1821          847           14 869
  ×  284        ×  319        ×  163        ×  6 134
   1 476        16 389        2 541         59 476
  29 520        18 210       50 820        446 070
  73 800       546 300       84 700      1 486 900
 104,796       580,899      138,061      89 214 000
                                         91,206,446
```

Try some exercises on multiplying large numbers.

Exercises

Work each of the following problems.

1. 567
   ×245

2. 3704
   × 286

3. 3964
   × 507

4. 416
   ×327

5. 564
   ×312

6. 362
   ×214

7. 206
   ×145

8. 524
   ×304

9. 663
   ×123

10. 3124
    ×1321

11. 448 × 260

12. 630 × 265

13. 4,682 × 833

14. 603 × 587

15. 6,827 × 373

16. After 10 years of work, a diesel locomotive was retired from service. It had made 1,564 transcontinental trips, each of which was 3,428 miles long. How many miles did it run altogether?

Example   Suppose a coast-to-coast television program costs $50,000 per minute to show on the air. What would a 2-hour movie cost?

$2 \times 60 = 120$ Change hours to minutes. 1 hour is 60 minutes, so 2 hours is 120 minutes.

$$\begin{array}{r} 50\ 000 \\ \times\qquad 120 \\ \hline 00\ 000 \\ 1\ 000\ 000 \\ 5\ 000\ 000 \\ \hline 6{,}000{,}000 \end{array}$$ Multiply $50,000 per minute by 120 minutes in all.

A 2-hour movie would cost $6,000,000 to show.

Example   In the spring, sap is collected from maple trees to make syrup. In a good season, a tree can produce 2 quarts of sap per hour. How much sap can be produced in 1 week?

$1 \times 7 = 7$ Change 1 week into days. 1 week is 7 days.

$7 \times 24 = 168$ Change days into hours. Multiply 7 days by 24 hours per day.

$$\begin{array}{r} ^{1\ 1}\\ 168 \\ \times\qquad 2 \\ \hline 336 \end{array}$$ Multiply 168 hours by 2 quarts of sap per hour.

A maple tree can produce 336 quarts of sap in 1 week.

Exercises

1. Mary's office club was raising money for an orphanage. The company management said that it would contribute 3 times whatever the club collected. Bill collected $142, Eleanor collected $139, and Mary collected $255. How much did the company contribute?

2. According to the encyclopedia, a hummingbird's wings beat 50 times per second. If it takes 3 minutes for a hummingbird to fly from its nest to a patch of wildflowers, how many times would its wings have to beat?

3. The average person has a heart rate of 72 beats per minute. How many times does the heart beat in 1 day (24 hours)?

4. A good racehorse can keep up an average speed of 45 miles per hour for a long time. If these racehorses had been available back in the days of the pony express, how far could a letter have traveled in 2 days?

5. We think of a light bulb as a steady source of light. Actually, it is flashing or blinking 60 times per second. If it takes 6 minutes to read a page like this one, how many times would a light bulb blink during that time?

Answers

1. $1,608   2. 9,000 times   3. 103,680 times   4. 2,160 miles   5. 21,600 times

Review Word Problem

For a break and a review, here's a chance to combine your knowledge of addition, subtraction, and multiplication to work a single problem. Work carefully and you should have fun!

George Stanley, a barber, charged $5 per head. If the customer was bald, he charged $2 per head plus $3 to hunt for the hair. The business for a typical week is shown below.

Monday — 12 heads + 3 bald ones
Tuesday — 9 heads + 2 bald ones
Wednesday — closed all day
Thursday — 16 heads + no bald ones
Friday and Friday evening — 18 heads of which 6 were bald
Saturday — 14 heads + no bald ones

George's expenses for the week were as follows:

Cleaning towels — $7
Building rent — $172
Axle grease — $13
Band-Aids — $27

What was George's profit for the week, if he used $5 of his money to get his own hair cut?

The solution is below, but DON'T LOOK until you have tried it.

Solution					
					PROFIT: $370
					− 224
					$146
INCOME:	Monday	15 × $5 = $ 75	EXPENSES:	Towels	$ 7
	Tuesday	11 × $5 = 55		Rent	172
	Thursday	16 × $5 = 80		Grease	13
	Friday	18 × $5 = 90		Band-Aids	27
	Saturday	14 × $5 = 70		Haircut	5
		Total $370		Total	$224

Unit 3 Test

Use a separate piece of paper to work out each problem.

Multiply.

1. 3×8

2. 7×6

3.
$$421 \times 3$$

4.
$$64 \times 9$$

5.
$$85 \times 7$$

6.
$$61 \times 7$$

7.
$$56 \times 6$$

8.
$$45 \times 18$$

9.
$$64 \times 53$$

10.
$$53 \times 94$$

11.
$$45 \times 72$$

12.
$$25 \times 48$$

13.
$$77 \times 42$$

14.
$$6304 \times 46$$

15.
$$407 \times 36$$

16.
$$6507 \times 34$$

17.
$$322144 \times 56$$

18.
$$677 \times 48$$

19.
$$625 \times 914$$

20.
$$957 \times 674$$

21.
$$673 \times 304$$

22. 875×20

23. $2{,}014 \times 859$

24. 563×217

25. If you save $27 a week, how much money will you have after 3 years? (1 year is 52 weeks.)

26. Shelly pays $385 per month to rent her apartment. How much rent would she pay for 3 years?

Your Answers

1. _____
2. _____
3. _____
4. _____
5. _____
6. _____
7. _____
8. _____
9. _____
10. _____
11. _____
12. _____
13. _____
14. _____
15. _____
16. _____
17. _____
18. _____
19. _____
20. _____
21. _____
22. _____
23. _____
24. _____
25. _____
26. _____

4 | Dividing Whole Numbers

Aims you toward:

- Dividing numbers and getting the right answer.
- Knowing how to check your answer to a division problem.

In this unit you will work exercises like these:

- Divide 147 by 7.
- Divide and check your answer: 4329 ÷ 60.

Dividing Numbers

We use DIVISION to solve a wide variety of everyday problems. It's important to DIVIDE things the right way. Suppose you had to split a case of 24 bottles of Dr Pepper among 4 students equally.

24 bottles

4 groups of 6 bottles

4 students

You divide 24 bottles among 4 students. Each student gets 6 bottles.

24 divided by 4 is 6.

$$24 \div 4 = 6$$

The ÷ sign means the first number "divided by" the second number. A division problem can also be written as $4\overline{)24}$. This means 4 "divided into" 24. Here you put the answer (quotient) on top, lined up above the 4 in the dividend.

(divisor) $4\overline{)24}$ (dividend) $\overset{6}{\phantom{4\overline{)2}}}$ (quotient)

To help you learn to divide, we first note that

DIVIDING is the OPPOSITE of MULTIPLYING.

Example   Compare these two statements.

If 4 students each had 6 bottles of Dr Pepper, they had 4 × 6 or 24 bottles altogether. If 24 bottles of Dr Pepper are divided among 4 students, they would each get 24 ÷ 4 or 6 bottles.

In your multiplication tables, you learned that 5 × 4 = 20.

Now you can also say that 20 ÷ 4 = 5, or $4\overline{)20}$ with 5 on top.

And 20 ÷ 5 = 4, or $5\overline{)20}$ with 4 on top.

Try the exercises on the next page.

Exercises

1. $6 \times 8 =$ _____   **2.** $6 \times$ _____ $= 48$ **3.** $48 \div 6 =$ _____   **4.** $6\overline{)48}$

5. $8 \times 6 =$ _____   **6.** $8 \times$ _____ $= 48$ **7.** $48 \div 8 =$ _____   **8.** $8\overline{)48}$

9. $9 \times 8 =$ _____   **10.** $9 \times$ _____ $= 72$ **11.** $72 \div 9 =$ _____   **12.** $9\overline{)72}$

13. $8 \times 9 =$ _____   **14.** $8 \times$ _____ $= 72$ **15.** $72 \div 8 =$ _____   **16.** $8\overline{)72}$

17. $5\overline{)30}$ **18.** $6\overline{)42}$ **19.** $8\overline{)32}$ **20.** $9\overline{)81}$ **21.** $8\overline{)72}$

22. $5\overline{)20}$ **23.** $8\overline{)48}$ **24.** $4\overline{)20}$ **25.** $5\overline{)40}$ **26.** $8\overline{)64}$

27. $8\overline{)40}$ **28.** $7\overline{)42}$ **29.** $9\overline{)54}$ **30.** $4\overline{)28}$ **31.** $6\overline{)30}$

32. $6\overline{)36}$ **33.** $7\overline{)28}$ **34.** $5\overline{)45}$ **35.** $5\overline{)25}$ **36.** $9\overline{)45}$

37. $6\overline{)48}$ **38.** $4\overline{)32}$ **39.** $4\overline{)16}$ **40.** $6\overline{)54}$ **41.** $7\overline{)63}$

42. Jennifer goes to Weight Watchers every week. She wants to lose 18 pounds in 9 weeks. How many pounds will she need to lose each week?

Answers

42. 2 pounds

			37. 8
	41. 9	**40.** 9	**39.** 4	**38.** 8	**37.** 8
36. 5	**35.** 5	**34.** 9	**33.** 4	**32.** 6	**31.** 5
30. 7	**29.** 6	**28.** 6	**27.** 5	**26.** 8	**25.** 8
24. 5	**23.** 6	**22.** 4	**21.** 9	**20.** 9	**19.** 4
18. 7	**17.** 6	**16.** 9	**15.** 9	**14.** 9	**13.** 72
12. 8	**11.** 8	**10.** 8	**9.** 72	**8.** 6	**7.** 6
6. 9	**5.** 48	**4.** 8	**3.** 8	**2.** 8	**1.** 48

Dividing with a Remainder

Sometimes it's impossible to divide something up evenly. Suppose you had to divide 13 coins into two EQUAL groups.

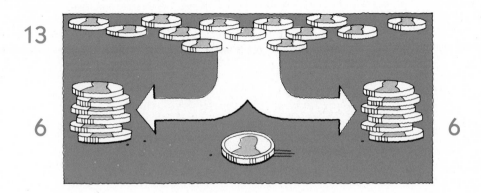

13

6   6

THERE'S ONE LEFT.

13 coins divided into 2 groups leaves 6 coins in each group and 1 coin left over.

$$13 \div 2 = 6, \text{ with a remainder of 1}$$

We work the problem like this:

$2\overline{)13}$ Ask yourself: $2 \times ?$ is almost 13.
         2×6 is almost 13, so we put a 6 above the 3.

$\begin{array}{r} 6 \\ 2\overline{)13} \\ 12 \end{array}$ Now multiply 2×6.
         $2 \times 6 = 12$, so put a 12 under the 13.

$\begin{array}{r} 6 \\ 2\overline{)13} \\ 12 \\ \hline 1 \end{array}$ Subtract.
         $13 - 12 = 1$
         The answer is 6, remainder 1.

Example   Divide 38 by 7.

$7\overline{)38}$ Be sure to set up the problem correctly.

$\begin{array}{r} 5 \\ 7\overline{)38} \end{array}$ Ask yourself: $7 \times ?$ is almost 38.
         7×5 is almost 38, so put a 5 above the 8.

$\begin{array}{r} 5 \\ 7\overline{)38} \\ 35 \\ \hline 3 \end{array}$ Multiply 7×5.
         Put the 35 under the 38 and subtract.
         The answer is 5, remainder 3.

Now do some exercises.

Exercises

1. $\begin{array}{r} 5 \\ 3\overline{)15} \\ \underline{15} \\ 0 \end{array}$

2. $\begin{array}{r} 5 \\ 3\overline{)16} \\ \underline{15} \\ 1 \end{array}$

3. $3\overline{)17}$

4. $4\overline{)19}$

5. $5\overline{)25}$

6. $5\overline{)26}$

7. $5\overline{)27}$

8. $5\overline{)28}$

9. $3\overline{)9}$

10. $3\overline{)10}$

11. $3\overline{)11}$

12. $3\overline{)12}$

13. $7\overline{)42}$

14. $7\overline{)43}$

15. $7\overline{)44}$

16. $7\overline{)45}$

17. $7\overline{)46}$

18. $7\overline{)47}$

19. $7\overline{)48}$

20. $7\overline{)49}$

21. $24 \div 6$

22. $26 \div 6$

23. $28 \div 6$

24. $30 \div 6$

25. Joanne has $27. How many scarves can she buy for $6 each? How much money will she have left?

26. A grocery store manager gave $84 to 4 children for carrying groceries to customers' cars. If the money was divided evenly between the children, how much did each child earn?

Answers

25. 4 scarves, $3 left		26. $21	
21. 4	22. 4 R2	23. 4 R4	24. 5
17. 6 R4	18. 6 R5	19. 6 R6	20. 7
13. 6	14. 6 R1	15. 6 R2	16. 6 R3
9. 3	10. 3 R1	11. 3 R2	12. 4
5. 5	6. 5 R1	7. 5 R2	8. 5 R3
1. 5	2. 5 R1	3. 5 R2	4. 4 R3

Larger Quotients

Suppose you want to divide 84 by 4.

$$4\overline{)84}$$

$$\begin{array}{r} 2 \\ 4\overline{)84} \\ 8 \\ \hline 0 \end{array}$$
First think of $4\overline{)8}$.
$4 \times 2 = 8$, so $8 \div 4 = 2$.

$$\begin{array}{r} 2 \\ 4\overline{)84} \\ 8 \\ \hline 04 \end{array}$$
Bring down the 4 from the dividend.

$$\begin{array}{r} 21 \\ 4\overline{)84} \\ 8 \\ \hline 04 \\ 4 \\ \hline 0 \end{array}$$
Now think of $4\overline{)4}$.
$4 \times 1 = 4$, so $4 \div 4 = 1$.
The answer is 21.

Example   Divide 3 into 97.

$$3\overline{)97}$$

$$\begin{array}{r} 3 \\ 3\overline{)97} \\ 9 \\ \hline 0 \end{array}$$
Think of $3\overline{)9}$.
$3 \times 3 = 9$, so $9 \div 3 = 3$.

$$\begin{array}{r} 3 \\ 3\overline{)97} \\ 9 \\ \hline 07 \end{array}$$
Bring down the 7.

$$\begin{array}{r} 32 \\ 3\overline{)97} \\ 9 \\ \hline 07 \\ 6 \\ \hline 1 \end{array}$$
Now think of $3\overline{)7}$.
$3 \times 2 = 6$, so $7 \div 3 = 2$, remainder 1.
The answer is 32, remainder 1.

Work the exercises on the next page.

Exercises

1. 5$\overline{)55}$ 2. 5$\overline{)58}$ 3. 3$\overline{)93}$ 4. 3$\overline{)95}$

5. 2$\overline{)48}$ 6. 2$\overline{)49}$ 7. 3$\overline{)69}$ 8. 3$\overline{)68}$

9.
$$\begin{array}{r} 21 \\ 5\overline{)105} \\ \underline{10} \\ 05 \\ \underline{5} \\ 0 \end{array}$$
Make sure the 2 is over the 0 and the 1 is over the 5.

10. 5$\overline{)106}$ 11. 7$\overline{)147}$

12. 7$\overline{)148}$ 13. 64 ÷ 3   14. 37 ÷ 3   15. 43 ÷ 2

16. 56 ÷ 5   17. 128 ÷ 6   18. 169 ÷ 8   19. 108 ÷ 5

20. Joe's paycheck showed 189 hours of work. He was a good worker, putting in 9 hours a day. How many days did he work?

21. How many weeks are there in 79 days? How many days are left over?

22. Four buses were loaded with 96 passengers on an excursion trip. How many people did each bus carry?

Answers

21. 11 weeks, 2 days

22. 24 people

20. 21 days   19. 21 R3   18. 21 R1   17. 21 R2

16. 11 R1   15. 21 R1   14. 12 R1   13. 21 R1

12. 21 R1   11. 21   10. 21 R1   9. 21

8. 22 R2   7. 23   6. 24 R1   5. 24

4. 31 R2   3. 31   2. 11 R3   1. 11

Suppose you want to divide 92 by 4.

$$4\overline{)92}$$

$$\begin{array}{r} 2 \\ 4\overline{)92} \\ 8 \\ \hline 1 \end{array}$$ Think of $4\overline{)9}$.
$4 \times 2 = 8$, so $9 \div 4 = 2$, remainder 1.

$$\begin{array}{r} 2 \\ 4\overline{)92} \\ 8 \\ \hline 12 \end{array}$$ Bring down the 2.
This makes 12 altogether.

$$\begin{array}{r} 23 \\ 4\overline{)92} \\ 8 \\ \hline 12 \\ 12 \\ \hline 0 \end{array}$$ Now think of $4\overline{)12}$.
$4 \times 3 = 12$, so $12 \div 4 = 3$.
The answer is 23.

Another Example   Divide 58 by 3.

$$3\overline{)58}$$

$$\begin{array}{r} 1 \\ 3\overline{)58} \\ 3 \\ \hline 2 \end{array}$$ Think of $3\overline{)5}$.

$$\begin{array}{r} 1 \\ 3\overline{)58} \\ 3 \\ \hline 28 \end{array}$$ Bring down the 8.

$$\begin{array}{r} 19 \\ 3\overline{)58} \\ 3 \\ \hline 28 \\ 27 \\ \hline 1 \end{array}$$ Now think of $3\overline{)28}$.
The answer is 19, remainder 1.

Exercises are next.

Exercises

1. 4)78
2. 4)79
3. 8)176
4. 8)178

5. 3)186
6. 3)187
7. 3)197
8. 6)257

9. 6)252
10. 4)195
11. 4)192
12. 4)189

13. 56 ÷ 4
14. 97 ÷ 8
15. 88 ÷ 5
16. 73 ÷ 4

17. 235 ÷ 8
18. 174 ÷ 2
19. 370 ÷ 4

20. Allan tried to lose weight by eating lettuce for lunch instead of fries. He improved the flavor by adding a lot of diet salad dressing. In fact, he went through 225 ounces in 5 months. How many ounces per month did he use?

21. Keith and Lynell Slaverio are planning a 3-month (92-day) trip to Europe. They want to visit 6 different countries. How many days can they spend in each country if they spend the same number of days in each? How many days will they have left to rest and recover?

22. A desert army unit was allotted 132 quarts of water per day. If each soldier received 3 quarts per day, how many soldiers were in the unit?

Answers

22. 44 soldiers
20. 45 ounces per month
21. 15 days in each country, 2 days left
19. 92 R2
18. 87
17. 29 R3
16. 18 R1
15. 17 R3
14. 12 R1
13. 14
12. 47 R1
11. 48
10. 48 R3
9. 42
8. 42 R5
7. 65 R2
6. 62 R1
5. 62
4. 22 R2
3. 22
2. 19 R3
1. 19 R2

If you're still a little confused about dividing, this page will be a good review. You'll be learning new skills, too.

Suppose you want to divide 4 into 953.

4)953

<pre>
  2
4)953   First think of 4)9.
  8
  1
</pre>

<pre>
  23
4)953   Now bring down the 5.
  8   Think of 4)15.
 15
 12
  3
</pre>

<pre>
  238
4)953   Bring down the 3.
  8   Think of 4)33.
 15   The answer is 238, remainder 1.
 12
  33
  32
   1
</pre>

Example   Divide 3,492 by 8.

8)3492

<pre>
   4
8)3492   8 is bigger than 3, so think of 8)34.
  32
   2
</pre>

<pre>
  43
8)3492   Bring down the 9.
  32   Think of 8)29.
  29
  24
   5
</pre>

<pre>
  436
8)3492   Bring down the 2.
  32   Think of 8)52.
  29   The answer is 436,
  24   remainder 4.
  52
  48
   4
</pre>

More exercises are on the next page.

Exercises

1. 6)978
2. 9)684
3. 8)456
4. 4)2631

5. 8)6994
6. 7)4618
7. 5)2490
8. 9)7998

9. 6)2736
10. 4)495
11. 8)3649
12. 2)3158

13. 4)758
14. 3)814
15. 3)2917
16. 8)6614

17. $471 \div 2$
18. $2,529 \div 8$
19. $544 \div 3$
20. $729 \div 6$

21. Bonnie operates a rabbitry (rabbit farm) in her spare time. She sells the rabbits wholesale to pet stores. Last Easter Bonnie received $492. If she sold each rabbit for $3, how many rabbits did she sell?

22. At a summer camp with an enrollment of 675, the campers were split up evenly into 9 groups for an overnight hike. How many campers were in each group?

Answers

22. 75 campers
21. 164 rabbits
20. 121 R3
19. 181 R1
18. 316 R1
17. 235 R1
16. 826 R6
15. 972 R1
14. 271 R1
13. 189 R2
12. 1,579
11. 456 R1
10. 123 R3
9. 456
8. 888 R6
7. 498
6. 659 R5
5. 874 R2
4. 657 R3
3. 57
2. 76
1. 163

Two-Digit Divisors

So far we've been dividing by a one-digit divisor. ⟶

$$\begin{array}{r} 18 \\ 4\overline{)75} \\ 4 \\ \hline 35 \\ 32 \\ \hline 3 \end{array}$$

Watch what happens when we use a two-digit divisor.

Divide 75 by 34.

$34\overline{)75}$ You want to know: 34 × ? is almost 75.
To make this easier, do some estimating.
34 is about 30.
75 is about 80.
Forget the zeros. Ask: 3 × ? is almost 8.

$$\begin{array}{r} 2 \\ 34\overline{)75} \end{array}$$
We know that 3 × 2 is almost 8.
Put a 2 above the 5 in the quotient.

$$\begin{array}{r} 2 \\ 34\overline{)75} \\ 68 \\ \hline 7 \end{array}$$
Multiply 2 × 34.
Subtract 75 − 68.
The answer is 2, remainder 7.

Another Example

Divide 85 by 31.

$31\overline{)85}$ 31 is about 30.
85 is about 90.
3 × ? is almost 9.

$$\begin{array}{r} 3 \\ 31\overline{)85} \end{array}$$
3 × 3 is 9.
Put a 3 above the 5.

$$\begin{array}{r} 3 \\ 31\overline{)85} \\ 93 \end{array}$$
Multiply 3 × 31.
Subtract 85 − 93. Can't do it.
So you know you've tried the wrong number.
(This happens sometimes when we estimate. Don't worry; just try the next smaller number.)

$$\begin{array}{r} 2 \\ 31\overline{)85} \\ 62 \\ \hline 23 \end{array}$$
Try 2.
Multiply 2 × 31.
Subtract 85 − 62.
The answer is 2, remainder 23.

Sometimes when we estimate, we choose a trial quotient that is too small. You know that this has happened if your remainder is larger than the divisor. So if you end up with a remainder that is larger than the divisor, rework the problem using the next higher number for the quotient.

Exercises

1. $41\overline{)88}$ 2. $53\overline{)60}$ 3. $81\overline{)92}$ 4. $21\overline{)58}$

5. $32\overline{)74}$ 6. $46\overline{)99}$ 7. $80\overline{)95}$ 8. $52\overline{)77}$

9. $24\overline{)96}$ 10. $33\overline{)87}$ 11. $38\overline{)94}$ 12. $18\overline{)67}$

13. $62 \div 13$ 14. $73 \div 29$ 15. $93 \div 31$ 16. $77 \div 20$

Answers

16. 3 R17   15. 3   14. 2 R15   13. 4 R10
12. 3 R13   11. 2 R18   10. 2 R21   9. 4
8. 1 R25   7. 1 R15   6. 2 R7   5. 2 R10
4. 2 R16   3. 1 R11   2. 1 R7   1. 2 R6

Suppose you want to divide 483 by 23.

23)483 Think of 23)48.
     23 is about 20, and 48 is about 50.
     Think: 20 × ? is almost 50. 2 × ? is almost 5.

  2 2 × 2 is almost 5.
23)483 Put a 2 in the quotient.
 46 Multiply 2 × 23.
  2 Subtract 48 − 46.

  2
23)483 Bring down the 3.
 46
 23

 21 Think of 23)23.
23)483 You know this is 1.
 46 Put a 1 in the quotient.
 23 Multiply 1 × 23.
 23 Subtract 23 − 23.
  0 The answer is 21.

Another Example Divide: 1,365 ÷ 46.

46)1365 Think of 46)136.
     46 is about 50, and 136 is about 140.
     Think: 50 × ? is almost 140. 5 × ? is almost 14.

  2 5 × 2 is almost 14.
46)1365 Put a 2 in the quotient.
 92 Multiply 2 × 46.
 44 Subtract 136 − 92.

 29 Bring down the 5.
46)1365 Think of 46)445.
 92 46 is about 50, and 445 is about 450.
 445 Think: 50 × ? is almost 450. 5 × ? is almost 45.
 414 5 × 9 is 45.
 31 Put a 9 in the quotient. Multiply 9 × 46.
     Subtract 445 − 414. The answer is 29, remainder 31.

Let's do some more exercises.

Exercises

1. 46)966 **2.** 28)644 **3.** 52)1625 **4.** 72)2536

5. 64)2304 **6.** 25)928 **7.** 27)649 **8.** 43)1691

9. 61)2930 **10.** 35)1509 **11.** 27)1137 **12.** 96)3938

13. 704 ÷ 21 **14.** 927 ÷ 62 **15.** 3,147 ÷ 42 **16.** 2,956 ÷ 17

17. Suppose you are a foreman for a company that produces heating and cooling appliances. You are in charge of 12 assembly lines. Altogether, the 12 lines produced 672 machine parts in 1 hour. How many parts should each line have produced in 1 hour?

18. The cartoon movie *Cinderella* required over 3,000,000 drawings by Disney artists, and took 150 weeks to produce. How many drawings were done per week?

Answers

17. 56 parts **18.** 20,000 drawings

1. 21 **2.** 23 **3.** 31 R13 **4.** 35 R16
5. 36 **6.** 37 R3 **7.** 24 R1 **8.** 39 R14
9. 48 R2 **10.** 43 R4 **11.** 42 R3 **12.** 41 R2
13. 33 R11 **14.** 14 R59 **15.** 74 R39 **16.** 173 R15

Larger Divisors

We've worked division problems with one-digit divisors and two-digit divisors. You use the same methods to work problems with three-digit and even larger divisors.

Example   Divide 9,198 by 396.

396)9198   First think of 396)919.
            396 is about 400.
            919 is about 900.
            Think: 4 × ? is almost 9.

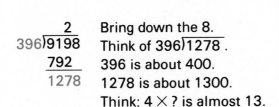

```
    2
396)9198
    792
    127
```
4 × 2 is almost 9.
Put a 2 in the quotient.
(Make sure you put the
2 in the correct place!)
Multiply 2 × 396.
Subtract 919 − 792.

```
    2
396)9198
    792
   1278
```
Bring down the 8.
Think of 396)1278 .
396 is about 400.
1278 is about 1300.
Think: 4 × ? is almost 13.

```
    23
396)9198
    792
   1278
   1188
     90
```
4 × 3 is almost 13.
Put a 3 in the quotient.
Multiply 3 × 396.
Subtract 1278 − 1188.
The answer is 23, remainder 90.

More Examples

```
        52
714)37218
    3570
    1518
    1428
      90
```

```
         341
2352)802114
     7056
     9651
     9408
     2434
     2352
       82
```

```
         133
3526)469473
     3526
    11687
    10578
    11093
    10578
      515
```

Now work the exercises.

Exercises

1. 437)2507

2. 402)15724

3. 315)927

4. 688)9215

5. 735)61394

6. 394)26530

7. 216)1195

8. 263)6553

9. 3174)39820

10. 9,215 ÷ 688

11. 9,247 ÷ 396

12. 6,553 ÷ 263

13. A woodlot has a stand of 175 fir trees. They produce 28,525 feet of lumber in one year. Assuming each tree produces the same amount, about how many feet does each tree produce?

14. In one day, the United Nations distributed 2,916 pounds of food to 643 orphaned children. How many pounds did each child get? How many pounds of food were left for the next morning?

Answers

14. 4 pounds; 344 pounds

13. 163 feet

10. 13 R271   **11.** 23 R139   **12.** 24 R241

7. 5 R115   **8.** 24 R241   **9.** 12 R1732

4. 13 R271   **5.** 83 R389   **6.** 67 R132

1. 5 R322   **2.** 39 R46   **3.** 2 R297

Zeros in the Quotient

Suppose you want to divide 3,654 by 9. (Aren't you glad we're back to one-digit divisors?!)

```
     4
9)3654        Think of 9)36.
  36
   0
```

```
     4
9)3654        Bring down the 5.
  36
  05
```

```
    40
9)3654        9 is bigger than 5.
  36          1 × 9 is too big.
  05          So use 0 × 9 = 0.
   0
   5
```

```
   406
9)3654        Now bring down the 4.
  36          Think of 9)54.
  05          The answer is 406.
   0
  54
  54
   0
```

Example   Divide 78 into 795.

```
    1
78)795        Think of 78)79.
   78
    1
```

```
    1
78)795        Bring down the 5.
   78
   15
```

```
   10
78)795        78 is bigger than 15.
   78         1 × 78 is too big.
   15         So use 0 × 78 = 0.
    0         The answer is 10,
   15         remainder 15.
```

Exercises

1. $7\overline{)2138}$

2. $5\overline{)3527}$

3. $8\overline{)16334}$

4. $56\overline{)6000}$

5. $21\overline{)637}$

6. $36\overline{)7418}$

7. $24\overline{)7205}$

8. $23\overline{)9430}$

9. $72\overline{)7915}$

10. $38\overline{)212883}$

11. $39\overline{)7891}$

12. $45\overline{)9307}$

Answers

1. 305 R3
2. 705 R2
3. 2,041 R6
4. 107 R8
5. 30 R7
6. 206 R2
7. 300 R5
8. 410
9. 109 R67
10. 5,602 R7
11. 202 R13
12. 206 R37

Checking Division Problems

Do you remember how we checked subtraction problems by adding? We can also check division problems by multiplying.

Example   Divide: 131 ÷ 27.

$$\begin{array}{r} 4 \\ 27\overline{)131} \\ \underline{108} \\ 23 \end{array}$$ The answer is 4, remainder 23.

Here's how to check your answer:

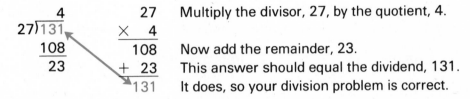

$$\begin{array}{r} 27 \\ \times\ \ 4 \\ \hline 108 \\ +\ \ 23 \\ \hline 131 \end{array}$$

Multiply the divisor, 27, by the quotient, 4.

Now add the remainder, 23.
This answer should equal the dividend, 131.
It does, so your division problem is correct.

Checking a Division Problem	Multiply the divisor and the quotient. Add the remainder. This answer should equal the dividend in the original problem.

Another Example   Divide 676 by 43 and check your answer.

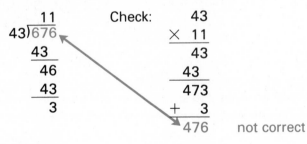

Check:

$$\begin{array}{r} 43 \\ \times\ 11 \\ \hline 43 \\ 43 \\ \hline 473 \\ +\ \ \ 3 \\ \hline 476 \end{array}$$ not correct

Try again.

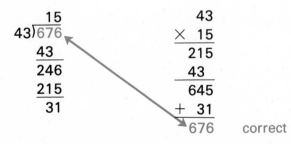

$$\begin{array}{r} 43 \\ \times\ 15 \\ \hline 215 \\ 43 \\ \hline 645 \\ +\ \ 31 \\ \hline 676 \end{array}$$ correct

Note that this method for checking division problems will not tell you exactly where you have made an error, only that you have made one.

Now work some exercises on checking division problems.

Exercises

Check the following division problems. Some are correct; some are not correct.

$$
\begin{array}{r}
7 \\
1.\ 79\overline{)627} \\
553 \\
\overline{74}
\end{array}
$$

$$
\begin{array}{r}
51 \\
2.\ 93\overline{)4753} \\
465 \\
\overline{103} \\
93 \\
\overline{10}
\end{array}
$$

$$
\begin{array}{r}
155 \\
3.\ 38\overline{)5820} \\
38 \\
\overline{202} \\
180 \\
\overline{220} \\
190 \\
\overline{30}
\end{array}
$$

Work the following division problems and check your answers.

4. $15\overline{)6428}$

5. $28\overline{)8160}$

6. $60\overline{)4329}$

7. $62\overline{)18943}$

8. $45\overline{)9307}$

9. $822\overline{)3762}$

10. $443\overline{)243976}$

11. $216\overline{)1195}$

12. $394\overline{)26530}$

13. Heather's father complained that she spent too much time on the telephone. He said that in one evening she used the phone 240 minutes. However, Heather said that this was only about 3 hours. Was she right?

14. A Holstein calf gained 375 pounds in 125 days. How many pounds did it gain per day?

Review Exercises

Work these problems to help prepare yourself for the next test. Check your answers by multiplying before you look at the answers below.

1. $9\overline{)89}$ **2.** $8\overline{)57}$ **3.** $2\overline{)85}$ **4.** $2\overline{)68}$

5. $7\overline{)238}$ **6.** $6\overline{)594}$ **7.** $3\overline{)340}$ **8.** $8\overline{)647}$

9. $11\overline{)689}$ **10.** $27\overline{)6997}$ **11.** $34\overline{)81610}$ **12.** $816\overline{)121517}$

13. Sheila VanNorman makes ceramic vases and sells them for $13 each. She sold some to an art store and received $117. How many did she sell?

14. G. R. Gumbull rented her lakeside cottage to Dave and Joy Scafe for 37 weeks. The Scafes paid $11,348 altogether. Joy figured that the rent came to more than $300 per week. Was she right?

Answers

1. 9 R8   **2.** 7 R1   **3.** 42 R1   **4.** 34   **5.** 34

6. 99   **7.** 113 R1   **8.** 80 R7   **9.** 62 R7   **10.** 259 R4

11. 2,400 R10   **12.** 148 R749   **13.** 9 vases   **14.** Yes

Review Word Problem

If you can figure out how to solve this problem and get the right answers, then you're doing very well. Take your time!

Suppose you are a post office worker. Your job is to make sure the daily mail is sorted properly and efficiently. Mail comes to your office from all over the world. In one day there are 76 letters from overseas, 1,935 letters from the East Coast, 3,720 letters from the West Coast, 2,113 letters from the Midwest, 2,399 letters from the South, and 2,705 letters from other regions of the country.

Your post office serves a town that has 3 zip codes, or postal codes, so the mail is first divided according to code. Each code region is further divided into 13 areas. Finally, there are 83 homes in each area. Assume that each home receives the same amount of mail.

How many pieces will go to each code region?
How many will go to each area?
How many will go to each home?

Solution

Add to find the total amount of mail for the day.

```
     76
  1,935
  3,720
  2,113
  2,399
  2,705
 12,948
```

Now divide.

```
pieces for        4316
each zip     3)12948
code region     12
                 09
                  9
                 04
                  3
                 18
                 18
                  0
```

```
pieces for        332
each area    13)4316
                 39
                 41
                 39
                 26
                 26
                  0
```

```
pieces          4
for each    83)332
home           332
                 0
```

Exercises

1. If an oil drilling rig averages a depth of 6 feet per hour and the crew works several shifts around the clock, how many days will it take to bring in a well at a depth of 6,480 feet?

2. Suppose the well in Exercise 1 brings in 350 barrels of oil in 1 day, worth a total of $10,150. Compare the price per barrel of this oil to the price of OPEC oil, which is $31 per barrel. How much cheaper is the price per barrel of this oil?

3. Cheryl and Shawna were excited about moving into their first apartment. The landlord said that if they wanted to put paint on the walls and lay a new carpet, he would pay them $8 per hour and subtract it from their first month's rent of $450. Cheryl worked 37 hours, since she was on vacation from her job. Shawna, who was working at her regular job, could only spare 18 hours for the work. How much did the two women have to pay for the rent?

4. If the author of this book saw a snake while walking in the woods, and the snake was traveling 3 miles per hour south, in what direction would the author travel, and at what speed?

Answers

1. 45 days   2. $2   3. $10   4. North, at about 1 million miles per hour!

Unit 4 Test

Use a separate piece of paper to work out each problem.

Divide. Check each answer by multiplying.

1. 33)87

2. 278)820

3. 53)238

4. 32)128652

5. 12)817

6. 412)6717

7. 4)28440

8. 78)795

9. 51)17562

10. 827)62103

11. 63)8194

12. 84)9634

13. 15)6428

14. 15,724 ÷ 402

15. 8)64077

16. 904 ÷ 342

17. John Houser does his laundry every 5 days. How many times does he do laundry in a year (365 days)?

18. Alice Diedricksen can type 47 words per minute. How long will it take her to type 1,175 words?

Your Answers

1. _____

2. _____

3. _____

4. _____

5. _____

6. _____

7. _____

8. _____

9. _____

10. _____

11. _____

12. _____

13. _____

14. _____

15. _____

16. _____

17. _____

18. _____

Review Test

Every once in a while we'll be having a REVIEW TEST on a group of units. This is the first one. It covers the first four units. Work each problem carefully, then check your answers.

1. 786×48

2. $392 - 46$

3. $34 + 314 + 9$

4. 82×16

5. $918 \div 6$

6. $6,172 \div 21$

7. $827 - 98$

8. $4,211 + 316$

9. 18×3

10. $98 \div 2$

11. $1,983 \div 14$

12. $6,224 \times 375$

Your Answers

1. _____

2. _____

3. _____

4. _____

5. _____

6. _____

7. _____

8. _____

9. _____

10. _____

11. _____

12. _____

13. _____

14. _____

_____

13. Joani Robinson works part time at a grocery store. She worked 4 hours on Monday, 3 hours on Tuesday, 8 hours on Wednesday, 0 hours on Thursday, and 5 hours each on Friday and Saturday. If she earns $3 an hour, how much money did she make?

14. Joani wants to buy some sweaters for $20 apiece. How many sweaters can she buy with what she earned in the week described above? How much money will she have left?

5 | Multiplying and Dividing Fractions

Aims you toward:

- Writing fractions and mixed numbers.
- Raising fractions to higher terms.
- Reducing fractions to lower terms.
- Multiplying fractions and getting the right answer.
- Dividing fractions and getting the right answer.

In this unit you will work exercises like these:

- Change $7\frac{5}{6}$ to an improper fraction.

- Change $\frac{16}{3}$ to a mixed number.

- Reduce $\frac{12}{16}$ to its lowest terms.

- Multiply $\frac{7}{8} \times 1\frac{3}{4}$.

- Divide 9 by $\frac{1}{2}$.

Fractions

Often in math you deal with things that are LESS THAN ONE.

Suppose you are an insurance sales representative assigned to a large district. Every spring it is your responsibility to contact each of your customers. Your list of customers is particularly long this year, and you are not able to contact them all in one week, as you have in past years. You decide to divide the list into 4 parts and contact 1 part each week.

In other words you would contact $\frac{1}{4}$ ("one fourth") of the customers each week.

By the end of the third week, you would have contacted $\frac{3}{4}$ ("three fourths") of the customers.

The fraction $\frac{1}{4}$ means that you have 1 part and there are 4 equal parts altogether. $\frac{1}{4}$ of the figure to the right is in color.

The fraction $\frac{3}{4}$ means that you have 3 parts and there are 4 equal parts altogether. $\frac{3}{4}$ of the figure to the right is in color.

$\frac{1}{4}$ and $\frac{3}{4}$ are PROPER FRACTIONS; they are less than one whole thing. The whole thing would be $\frac{4}{4}$ or 1. $\frac{4}{4}$ of the figure to the right is in color.

Take a look at another example of a proper fraction.

Example   If you sleep 8 hours every day, what fraction of each full day do you sleep?

You sleep 8 hours (parts).
There are 24 hours (total parts) in a day.

So you sleep $\frac{8}{24}$ ("eight twenty-fourths") of each day.
This is the same as $\frac{1}{3}$ ("one third") of each day.

Equivalent Fractions

How did that go? Let's do another take on that idea.

Look at the solid lines in the box below. 2 parts are in color, and there are 3 parts total.

Or, $\frac{2}{3}$ of the box is in color.

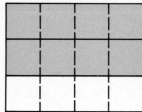

Now look at the dotted lines.
8 parts are in color, and there are 12 parts total.

$\frac{8}{12}$ of the box is in color.

In other words, $\frac{2}{3}$ is the same as $\frac{8}{12}$. They are called EQUIVALENT FRACTIONS, or equal fractions.

There is another way to see that $\frac{2}{3}$ is equivalent to $\frac{8}{12}$.

If you multiply the top (numerator) and the bottom (denominator) of a fraction by the same number, you get an equivalent fraction.

$$\text{(numerator)} \quad \frac{2 \times 4}{3 \times 4} = \frac{8}{12} \quad \text{(denominator)}$$

You can get an equivalent fraction for any given fraction in the same way.

Examples $\quad \frac{3 \times 3}{4 \times 3} = \frac{9}{12} \qquad \frac{3 \times 2}{4 \times 2} = \frac{6}{8} \qquad \frac{1 \times 5}{7 \times 5} = \frac{5}{35} \qquad \frac{5 \times 8}{8 \times 8} = \frac{40}{64}$

Sometimes when you want an equivalent fraction, you need a particular denominator or numerator.

Example Change $\frac{7}{8}$ to an equivalent fraction with 24 as the denominator.

$\frac{7}{8} = \frac{?}{24}$ $\qquad$ Set up the problem like this.

$\frac{7}{8 \times 3} = \frac{?}{24}$ $\qquad$ First ask $8 \times ? = 24$.
$\qquad\qquad\qquad$ You know that $8 \times 3 = 24$.

$\frac{7 \times 3}{8 \times 3} = \frac{21}{24}$ $\qquad$ Since the denominator was multiplied by 3, you must multiply the numerator by 3. The fraction equivalent to $\frac{7}{8}$ with denominator 24 is $\frac{21}{24}$.

This process is called RAISING A FRACTION TO HIGHER TERMS.

Example

Change $\dfrac{4}{5}$ to an equivalent fraction with 12 as the numerator.

$\dfrac{4}{5} = \dfrac{12}{?}$ Set up the problem like this.

$\dfrac{4}{5} = \dfrac{12}{?}$ What happened to the 4 to make it become 12?

$\dfrac{4 \times 3}{5} = \dfrac{12}{?}$ It must have been multiplied by 3, because $4 \times 3 = 12$.

$\dfrac{4 \times 3}{5 \times 3} = \dfrac{12}{?}$ If the numerator was multiplied by 3, then it is only fair that the denominator also be multiplied by 3.

$\dfrac{4 \times 3}{5 \times 3} = \dfrac{12}{15}$ Your missing denominator is 15.

Exercises

Fill in the missing numbers to make equivalent fractions. The first two are done to show you how.

1. $\dfrac{4}{9} = \dfrac{20}{?}$

2. $\dfrac{6}{11} = \dfrac{?}{22}$

3. $\dfrac{5}{8} = \dfrac{}{16}$

$\dfrac{4 \times 5}{9 \times 5} = \dfrac{20}{45}$

$\dfrac{6 \times 2}{11 \times 2} = \dfrac{12}{22}$

4. $\dfrac{7}{10} = \dfrac{14}{}$

5. $\dfrac{8}{9} = \dfrac{}{27}$

6. $\dfrac{4}{5} = \dfrac{20}{}$

7. $\dfrac{1}{3} = \dfrac{}{12}$

8. $\dfrac{1}{2} = \dfrac{5}{}$

9. $\dfrac{3}{11} = \dfrac{}{44}$

Answers

1. 45	2. 12	3. 10	4. 20	5. 24
6. 25	7. 4	8. 10	9. 12	

How about some more practice? Turn to the next page. ▶

Exercises

Fill in the missing numbers to make equivalent fractions.

1. $\dfrac{4}{5} = \dfrac{}{20}$

2. $\dfrac{5}{9} = \dfrac{}{36}$

3. $\dfrac{3}{7} = \dfrac{6}{}$

4. $\dfrac{4}{7} = \dfrac{}{28}$

5. $\dfrac{7}{8} = \dfrac{21}{}$

6. $\dfrac{6}{11} = \dfrac{}{77}$

7. $\dfrac{1}{2} = \dfrac{}{64}$

8. $\dfrac{2}{3} = \dfrac{}{24}$

9. $\dfrac{5}{6} = \dfrac{}{48}$

10. $\dfrac{7}{8} = \dfrac{}{56}$

11. $\dfrac{9}{10} = \dfrac{}{60}$

12. $\dfrac{2}{7} = \dfrac{14}{}$

13. $\dfrac{7}{9} = \dfrac{}{36}$

14. $\dfrac{1}{8} = \dfrac{}{24}$

15. $\dfrac{3}{10} = \dfrac{15}{}$

16. $\dfrac{3}{5} = \dfrac{12}{}$

17. $\dfrac{7}{9} = \dfrac{}{27}$

18. $\dfrac{3}{8} = \dfrac{12}{}$

19. $\dfrac{1}{5} = \dfrac{}{20}$

20. $\dfrac{13}{14} = \dfrac{}{42}$

21. $\dfrac{5}{16} = \dfrac{25}{}$

22. $\dfrac{3}{32} = \dfrac{9}{}$

23. $\dfrac{11}{24} = \dfrac{}{48}$

24. $\dfrac{14}{15} = \dfrac{70}{}$

So far you have been raising fractions to higher terms. Sometimes you are asked to REDUCE A FRACTION TO LOWER TERMS. To REDUCE a fraction, find a number that will divide exactly into both the numerator and the denominator. This gives an equivalent (equal) fraction in lower terms.

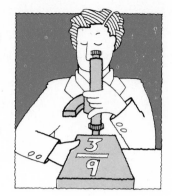

$$\frac{6}{18} \div \frac{2}{2} = \frac{3}{9}$$

$\frac{3}{9}$ is equivalent to $\frac{6}{18}$, and is in lower terms.

Often you are asked to reduce a fraction to its LOWEST terms. You will end up with a fraction whose numerator and denominator cannot be divided exactly by the same number.

Example $\frac{3}{9}$ can be reduced some more. Find a number that will divide exactly into both the numerator and the denominator.

$$\frac{3 \div 3}{9 \div 3} = \frac{1}{3}$$

Example   Reduce $\frac{14}{42}$ to its lowest terms.

$\dfrac{14 \div 2}{42 \div 2} = \dfrac{7}{21}$ Find a number that will divide exactly into both 14 and 42. Is the fraction $\frac{7}{21}$ now in lowest terms?

No. Both 7 and 21 can be divided by 7.

$\dfrac{7 \div 7}{21 \div 7} = \dfrac{1}{3}$ Is this fraction in lowest terms? Yes. 1 and 3 cannot be divided exactly by the same number.

$\frac{14}{42}$ reduced to lowest terms is $\frac{1}{3}$.

There are other ways to work this same problem.

For instance: $\dfrac{14 \div 7}{42 \div 7} = \dfrac{2}{6}$ ▶ $\dfrac{2 \div 2}{6 \div 2} = \dfrac{1}{3}$

or $\dfrac{14 \div 14}{42 \div 14} = \dfrac{1}{3}$.

Any way is okay, just so your answer is $\frac{1}{3}$.

Exercises

Fill in the missing number to make equivalent fractions.

1. $\dfrac{60}{180} = \dfrac{}{90}$

2. $\dfrac{28}{42} = \dfrac{}{6}$

3. $\dfrac{15}{80} = \dfrac{3}{}$

4. $\dfrac{4}{32} = \dfrac{1}{}$

$\dfrac{60}{180} \div 2 = \dfrac{?}{90}$

Write each fraction in its lowest terms.

5. $\dfrac{8}{12}$

6. $\dfrac{2}{4}$

7. $\dfrac{7}{21}$

8. $\dfrac{5}{15}$

9. $\dfrac{3}{18}$

10. $\dfrac{2}{12}$

11. $\dfrac{7}{35}$

12. $\dfrac{6}{9}$

13. $\dfrac{3}{24}$

14. $\dfrac{8}{16}$

15. $\dfrac{6}{42}$

16. $\dfrac{14}{40}$

17. $\dfrac{48}{64}$

18. $\dfrac{5}{15}$

19. $\dfrac{15}{45}$

Sometimes the same number will divide exactly into the numerator and denominator of a fraction several times.

Example   Reduce $\dfrac{125}{325}$ to lowest terms.

One way: $\dfrac{125 \div 5}{325 \div 5} = \dfrac{25}{65}$ $\dfrac{25 \div 5}{65 \div 5} = \dfrac{5}{13}$

Another way: $\dfrac{125 \div 25}{325 \div 25} = \dfrac{5}{13}$

Here are some more exercises on reducing fractions.

Exercises

Write each fraction in lowest terms.

1. $\dfrac{10}{25}$ 2. $\dfrac{15}{20}$ 3. $\dfrac{15}{40}$ 4. $\dfrac{25}{50}$ 5. $\dfrac{8}{24}$

6. $\dfrac{14}{21}$ 7. $\dfrac{75}{125}$ 8. $\dfrac{21}{30}$ 9. $\dfrac{12}{24}$ 10. $\dfrac{18}{30}$

11. $\dfrac{20}{45}$ 12. $\dfrac{12}{18}$ 13. $\dfrac{10}{45}$ 14. $\dfrac{9}{21}$ 15. $\dfrac{18}{20}$

16. $\dfrac{13}{26}$ 17. $\dfrac{4}{14}$ 18. $\dfrac{50}{75}$ 19. $\dfrac{9}{30}$ 20. $\dfrac{16}{32}$

Answers

1. $\dfrac{2}{5}$ 2. $\dfrac{3}{4}$ 3. $\dfrac{3}{8}$ 4. $\dfrac{1}{2}$ 5. $\dfrac{1}{3}$ 6. $\dfrac{2}{3}$

7. $\dfrac{3}{5}$ 8. $\dfrac{7}{10}$ 9. $\dfrac{1}{2}$ 10. $\dfrac{3}{5}$ 11. $\dfrac{4}{9}$ 12. $\dfrac{2}{3}$

13. $\dfrac{2}{9}$ 14. $\dfrac{3}{7}$ 15. $\dfrac{9}{10}$ 16. $\dfrac{1}{2}$ 17. $\dfrac{2}{7}$ 18. $\dfrac{2}{3}$

19. $\dfrac{3}{10}$ 20. $\dfrac{1}{2}$

Mixed Numbers and Improper Fractions

Until now, you have been working with PROPER FRACTIONS (like $\frac{3}{4}$). Now come MIXED NUMBERS. A mixed number is a whole number and a fraction. An example of a mixed number is $4\frac{2}{25}$.

Take a look at these circles.

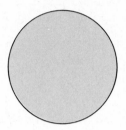  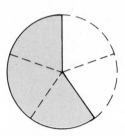

There are 2 whole circles plus $\frac{3}{5}$ of a third circle.

That is $2 + \frac{3}{5}$ circles. Leave out the plus sign and write $2\frac{3}{5}$.

$2\frac{3}{5}$ is the mixed number which represents the circles above.

The last type of fraction you need to know about is the IMPROPER FRACTION.

Look at this rectangle.

Now look at $\frac{1}{4}$ of the rectangle.

What happens if you put 9 of the $\frac{1}{4}$ portions together?

You have $\frac{9}{4}$ of the original rectangle.

$\frac{9}{4}$ is an improper fraction.

Remember:

A plain fraction like $\frac{3}{4}$ is called a PROPER FRACTION or a COMMON FRACTION.

A number like $2\frac{1}{2}$ is called a MIXED NUMBER.

A fraction with the larger number on top, like $\frac{12}{3}$, is called an IMPROPER FRACTION.

When you have an improper fraction, you often need to change it to a whole number or a mixed number. This is easy to do. Divide the numerator by the denominator.

For example, to change $\frac{12}{3}$ to a whole number or mixed number, divide 12 by 3.

$$\begin{array}{r} 4 \\ 3\overline{)12} \\ \underline{12} \\ 0 \end{array}$$ The answer is 4. $\frac{12}{3}$ is the same as 4.

Let's go through this again. First we should know that $\frac{12}{3}$ is the same as $12 \div 3$ or $3\overline{)12}$.

These are three ways of writing the same thing, and they all mean division.

So to change $\frac{12}{3}$ to a whole number or mixed number, you divide. $3\overline{)12}^{\,4}$

Example   Change $\frac{16}{4}$ to a whole number or mixed number.

$$\begin{array}{r} 4 \\ 4\overline{)16} \\ \underline{16} \\ 0 \end{array}$$ The answer is 4.

Some improper fractions will become mixed numbers.

Example   Change $\frac{7}{3}$ to a whole number or mixed number.

$$\begin{array}{r} 2 \\ 3\overline{)7} \\ \underline{6} \\ 1 \end{array}$$ You have a remainder of 1.
Put the remainder 1 over the divisor 3. This gives you $\frac{1}{3}$.
The final answer is $2\frac{1}{3}$.

To Change an Improper Fraction	Divide the numerator (top) by the denominator (bottom). If there is a remainder, put it over the denominator.

Whenever you divide and end up with a remainder, you can write your answer as a mixed number.

Example   Change $\frac{23}{4}$ to a whole number or mixed number.

$$\begin{array}{r} 5 \\ 4\overline{)23} \\ \underline{20} \\ 3 \end{array}$$ The answer is $5\frac{3}{4}$.

Exercises

For each problem, write a mixed number for the part in color.

1.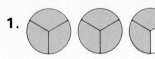
2.
3.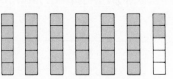

Name each of the following types of fractions.

4. $2\frac{3}{4}$ 5. $\frac{11}{4}$ 6. $\frac{24}{16}$ 7. $\frac{7}{8}$ 8. $6\frac{1}{3}$

Change the following improper fractions to whole numbers or mixed numbers.

9. $\frac{9}{2}$ 10. $\frac{10}{4}$ 11. $\frac{7}{3}$ 12. $\frac{60}{7}$ 13. $\frac{12}{5}$

14. $\frac{11}{2}$ 15. $\frac{8}{3}$ 16. $\frac{21}{3}$ 17. $\frac{63}{5}$ 18. $\frac{35}{7}$

19. Divide 107 by 4. Write your answer as a mixed number.

Answers

16. 7   17. $12\frac{3}{5}$ 18. 5   19. $26\frac{3}{4}$

11. $2\frac{1}{3}$ 12. $8\frac{4}{7}$ 13. $2\frac{2}{5}$ 14. $5\frac{1}{2}$ 15. $2\frac{2}{3}$

8. Mixed number   9. $4\frac{1}{2}$ 10. $2\frac{2}{4} = 2\frac{1}{2}$

5. Improper fraction   6. Improper fraction   7. Proper fraction

1. $2\frac{2}{3}$ 2. $2\frac{1}{4}$ 3. $6\frac{2}{5}$ 4. Mixed number

You just learned to change improper fractions to whole numbers or mixed numbers. Now you are going to learn to do it the other way around.

You know how to change $\frac{14}{3}$ to $4\frac{2}{3}$.

$$\begin{array}{r} 4 \\ 3\overline{)14} \\ 12 \\ \hline 2 \end{array}$$

To change $4\frac{2}{3}$ back to an improper fraction,

$$4 \, \diagdown \, \frac{2}{3} \diagdown$$

first multiply 4×3 and add on the 2.

$$4 \times 3 + 2 = 14$$

Put this answer over the denominator, 3.

$4\frac{2}{3}$ is the same as $\frac{14}{3}$.

$$\frac{14}{3}$$

To Change a Mixed Number to a Fraction	Multiply the whole number by the denominator. Add the numerator. Put the result over the denominator.

Example Change $3\frac{1}{4}$ to an improper fraction.

$$3 \times 4 = 12$$
$$12 + 1 = 13$$

Put the 13 over the 4, giving you $\frac{13}{4}$.

Example Change $7\frac{1}{2}$ to an improper fraction.

$$7 \times 2 = 14$$
$$14 + 1 = 15$$

Put the 15 over the 2, giving you $\frac{15}{2}$.

Exercises

Change the following mixed numbers to improper fractions.

1. $3\frac{2}{3}$ **2.** $5\frac{3}{4}$ **3.** $5\frac{5}{9}$ **4.** $14\frac{1}{3}$ **5.** $1\frac{2}{3}$

Fill in the blanks with either a mixed number or an improper fraction.

	Mixed number	Improper fraction		Mixed number	Improper fraction
6.	$7\frac{4}{9}$		**13.**	$7\frac{7}{8}$	
7.		$\frac{62}{9}$	**14.**		$\frac{14}{5}$
8.	$10\frac{1}{10}$		**15.**	$4\frac{7}{11}$	
9.		$\frac{39}{5}$	**16.**		$\frac{114}{8}$
10.	$7\frac{3}{7}$		**17.**	$36\frac{2}{3}$	
11.		$\frac{25}{3}$	**18.**		$\frac{19}{6}$
12.	$6\frac{2}{5}$		**19.**		$\frac{32}{3}$

Multiplying Fractions

Since the title of this unit is "Multiplying and Dividing Fractions," it's about time we got to it! If you are sure of everything so far, go on to MULTIPLYING FRACTIONS.

Let's start by multiplying $\frac{4}{5} \times \frac{2}{3}$.

$$\frac{4}{5} \times \frac{2}{3} = \frac{4 \times 2}{5 \times 3} = \frac{8}{15}$$

To Multiply Fractions	Multiply the numerators to form a new numerator. Multiply the denominators to form a new denominator. Reduce if needed.

Example Multiply $\frac{5}{6} \times \frac{3}{5}$.

$\frac{5}{6} \times \frac{3}{5} = \frac{5 \times 3}{6 \times 5} = \frac{15}{30}$ Multiply the numerators, 5×3. Then multiply the denominators, 6×5. You get $\frac{15}{30}$.

$\frac{15 \div 15}{30 \div 15} = \frac{1}{2}$ Always reduce your answer to LOWEST terms. The answer is $\frac{1}{2}$.

Some people like to use a shortcut method called CANCELING.

Example Multiply $\frac{5}{6} \times \frac{9}{10}$.

$\frac{5}{6} \times \frac{9}{10}$ To cancel, look for a number that will divide evenly into a numerator AND a denominator. 5 will divide evenly into both 5 (a numerator) and 10 (a denominator).

$\frac{\overset{1}{\cancel{5}}}{6} \times \frac{9}{\underset{2}{\cancel{10}}}$ $5 \div 5 = 1$ Cross out the 5 and replace it with a 1.
$10 \div 5 = 2$ Cross out the 10 and replace it with a 2.

$\frac{\overset{1}{\cancel{5}}}{\underset{2}{\cancel{6}}} \times \frac{\overset{3}{\cancel{9}}}{\underset{2}{\cancel{10}}}$ Now look at the 9 and the 6. 3 will divide evenly into both of them.
$9 \div 3 = 3$ $6 \div 3 = 2$

$\frac{1 \times 3}{2 \times 2} = \frac{3}{4}$ Now multiply. The answer is $\frac{3}{4}$.

Remember: You can cancel ONLY a numerator and denominator combination.
You can cancel ONLY when multiplying.

Exercises

Multiply the following. Cancel when you can. Make sure your answers are in their lowest terms.

1. $\dfrac{1}{2} \times \dfrac{3}{8}$ **2.** $\dfrac{2}{3} \times \dfrac{4}{5}$ **3.** $\dfrac{1}{8} \times \dfrac{3}{4}$ **4.** $\dfrac{5}{6} \times \dfrac{1}{3}$

5. $\dfrac{3}{4} \times \dfrac{5}{8}$ **6.** $\dfrac{3}{\underset{2}{8}} \times \dfrac{\overset{1}{4}}{5}$ **7.** $\dfrac{5}{8} \times \dfrac{9}{10}$ **8.** $\dfrac{2}{7} \times \dfrac{14}{15}$

9. $\dfrac{5}{8} \times \dfrac{7}{10}$ **10.** $\dfrac{5}{8} \times \dfrac{4}{7}$ **11.** $\dfrac{7}{16} \times \dfrac{1}{3}$ **12.** $\dfrac{9}{10} \times \dfrac{1}{6}$

13. $\dfrac{7}{12} \times \dfrac{3}{4}$ **14.** $\dfrac{9}{10} \times \dfrac{1}{3}$ **15.** $\dfrac{1}{4} \times \dfrac{11}{12}$ **16.** $\dfrac{3}{16} \times \dfrac{16}{3}$

Answers

1. $\dfrac{3}{16}$ **2.** $\dfrac{8}{15}$ **3.** $\dfrac{3}{32}$ **4.** $\dfrac{5}{18}$ **5.** $\dfrac{15}{32}$ **6.** $\dfrac{3}{10}$

7. $\dfrac{9}{16}$ **8.** $\dfrac{4}{15}$ **9.** $\dfrac{7}{16}$ **10.** $\dfrac{5}{14}$ **11.** $\dfrac{7}{48}$ **12.** $\dfrac{3}{20}$

13. $\dfrac{7}{16}$ **14.** $\dfrac{3}{10}$ **15.** $\dfrac{11}{48}$ **16.** 1

Sometimes you can cancel or reduce another way.

Examples   Multiply the following numbers.

$$\frac{4}{8} \times \frac{3}{6} \qquad\qquad \frac{5}{15} \times \frac{3}{9}$$

$$\frac{\overset{1}{\cancel{4}}}{\underset{2}{\cancel{8}}} \times \frac{\overset{1}{\cancel{3}}}{\underset{2}{\cancel{6}}} \qquad\qquad \frac{\overset{1}{\cancel{5}}}{\underset{3}{\cancel{15}}} \times \frac{\overset{1}{\cancel{3}}}{\underset{3}{\cancel{9}}}$$

$$\frac{1}{2} \times \frac{1}{2} = \frac{1}{4} \qquad\qquad \frac{1}{3} \times \frac{1}{3} = \frac{1}{9}$$

Exercises

Try these. The first one is worked out for you.

1. $\dfrac{\overset{1}{\cancel{5}}}{\underset{2}{\cancel{10}}} \times \dfrac{\overset{1}{\cancel{4}}}{\underset{3}{\cancel{12}}}$ 2. $\dfrac{2}{8} \times \dfrac{5}{25}$ 3. $\dfrac{2}{4} \times \dfrac{7}{14}$ 4. $\dfrac{5}{15} \times \dfrac{8}{14}$

$\dfrac{1}{2} \times \dfrac{1}{3} = \dfrac{1}{6}$

5. $\dfrac{8}{16} \times \dfrac{3}{9}$ 6. $\dfrac{10}{25} \times \dfrac{2}{6}$ 7. $\dfrac{3}{12} \times \dfrac{4}{12}$

8. $\dfrac{6}{12} \times \dfrac{15}{25}$ 9. $\dfrac{6}{8} \times \dfrac{3}{6}$ 10. $\dfrac{14}{21} \times \dfrac{8}{10}$

Answers

6. $\dfrac{2}{15}$ 7. $\dfrac{1}{12}$ 8. $\dfrac{3}{10}$ 9. $\dfrac{3}{8}$ 10. $\dfrac{8}{15}$

1. $\dfrac{1}{6}$ 2. $\dfrac{1}{20}$ 3. $\dfrac{1}{4}$ 4. $\dfrac{4}{21}$ 5. $\dfrac{1}{6}$

What happens if you want to multiply a fraction and a whole number, say $\frac{3}{4} \times 5$? First there's something you need to know.

Any number can be placed over 1 without changing the number.

$$5 = \frac{5}{1}$$

This makes sense because $\frac{5}{1}$ is the same as $5 \div 1$, and you know that $5 \div 1 = 5$.

Back to our problem. You multiply $\frac{3}{4} \times 5$ like this:

$$\frac{3}{4} \times 5 = \frac{3}{4} \times \frac{5}{1} = \frac{3 \times 5}{4 \times 1} = \frac{15}{4}$$

Once you've changed 5 to $\frac{5}{1}$, you can multiply easily.

$$\frac{15}{4} = 3\frac{3}{4}$$

For your final answer, always change an improper fraction to a whole number or a mixed number.

The answer is $3\frac{3}{4}$.

How do you multiply a fraction and a mixed number?

Example   Multiply $\frac{3}{4} \times 5\frac{1}{3}$.

$$5\frac{1}{3} = \frac{16}{3}$$

First change the mixed number to an improper fraction.

$$\overset{1}{\cancel{\underset{1}{\frac{3}{4}}}} \times \overset{4}{\cancel{\underset{1}{\frac{16}{3}}}} = \frac{1 \times 4}{1 \times 1} = \frac{4}{1}$$

Now multiply as you normally do. Canceling makes this problem easier.

$$\frac{4}{1} = 4$$

Remember to change an improper fraction to a whole number or a mixed number. The answer is 4.

To Multiply Mixed Numbers	Change the mixed numbers to improper fractions. Multiply as usual. (You do the same thing when you multiply a mixed number and a whole number.)

Now try some exercises.

Exercises

Multiply. If your answer is an improper fraction, change it to a whole number or a mixed number.

1. $\dfrac{5}{7} \times \dfrac{7}{10}$

2. $3\dfrac{1}{4} \times 1\dfrac{2}{4}$

$\dfrac{13}{4} \times \dfrac{6}{4}$

3. $5 \times \dfrac{3}{10}$

4. $2\dfrac{2}{3} \times 1\dfrac{1}{5}$

5. $7\dfrac{1}{2} \times \dfrac{4}{15}$

6. $\dfrac{6}{7} \times 2\dfrac{1}{3}$

7. $\dfrac{3}{4} \times 16$

8. $5 \times \dfrac{2}{5}$

9. $\dfrac{9}{24} \times \dfrac{12}{18}$

10. $4\dfrac{1}{3} \times \dfrac{3}{7}$

11. $2\dfrac{6}{8} \times 1\dfrac{4}{6}$

12. $9 \times \dfrac{11}{12}$

13. $1\dfrac{1}{2} \times 2\dfrac{3}{4}$

14. $3\dfrac{1}{4} \times 12$

15. $3\dfrac{2}{3} \times 15$

16. $1\dfrac{2}{5} \times \dfrac{5}{8}$

Answers

1. $\dfrac{1}{2}$ **2.** $4\dfrac{7}{8}$ **3.** $1\dfrac{1}{2}$ **4.** $3\dfrac{1}{5}$ **5.** 2 **6.** 2

7. 12 **8.** 2 **9.** $\dfrac{1}{4}$ **10.** $1\dfrac{6}{7}$ **11.** $4\dfrac{7}{12}$ **12.** $8\dfrac{1}{4}$

13. $4\dfrac{1}{8}$ **14.** 39 **15.** 55 **16.** $\dfrac{7}{8}$

Reciprocals: Getting Ready to Divide

Every number except 0 has a RECIPROCAL. A reciprocal of a number multiplied by the number itself is always 1. The idea is best explained by some examples.

The reciprocal of $\frac{2}{5}$ is $\frac{5}{2}$. $\frac{\cancel{2}^{1}}{\cancel{5}_{1}} \times \frac{\cancel{5}^{1}}{\cancel{2}_{1}} = \frac{1}{1} = 1$

The reciprocal of 7 is $\frac{1}{7}$. $7 \times \frac{1}{7} = \frac{7}{1} \times \frac{1}{\cancel{7}} = \frac{1}{1} = 1$

The reciprocal of $3\frac{1}{4}$ is $\frac{4}{13}$. $3\frac{1}{4} \times \frac{4}{13} = \frac{\cancel{13}^{1}}{\cancel{4}_{1}} \times \frac{\cancel{4}^{1}}{\cancel{13}_{1}} = \frac{1}{1} = 1$

To Find the Reciprocal of a Number	If the number is a whole number or a mixed number, change it to an improper fraction. Turn this fraction upside down. (This is called INVERTING a fraction.)

These two steps will give you the reciprocal of any number.

Example   What is the reciprocal of $7\frac{1}{4}$?

$7\frac{1}{4} = \frac{29}{4}$ First change the mixed number to an improper fraction.

$\frac{4}{29}$ Then invert the fraction.
The reciprocal of $7\frac{1}{4}$ is $\frac{4}{29}$.

Example   What is the reciprocal of 19?

$19 = \frac{19}{1}$ Change 19 to an improper fraction.

$\frac{1}{19}$ Invert the fraction.
The reciprocal of 19 is $\frac{1}{19}$.

Example   What is the reciprocal of $\frac{4}{5}$?

$\frac{5}{4}$ Since $\frac{4}{5}$ is already a fraction, all you need to do is invert it.
The reciprocal of $\frac{4}{5}$ is $\frac{5}{4}$.

Exercises

Give the reciprocal of each number.

1. 12

2. $\frac{1}{9}$

3. $2\frac{1}{2}$

4. 6

5. $3\frac{5}{8}$

6. $\frac{4}{7}$

7. 52

8. $\frac{7}{100}$

9. $4\frac{2}{3}$

10. $\frac{9}{10}$

11. $4\frac{6}{7}$

12. $10\frac{1}{3}$

13. $\frac{7}{16}$

14. $1\frac{2}{9}$

15. $3\frac{5}{8}$

Answers

1. $\frac{1}{12}$

2. 9

3. $\frac{2}{5}$

4. $\frac{1}{6}$

5. $\frac{8}{29}$

6. $\frac{7}{4}$

7. $\frac{1}{52}$

8. $\frac{100}{7}$

9. $\frac{3}{14}$

10. $\frac{10}{9}$

11. $\frac{7}{34}$

12. $\frac{3}{31}$

13. $\frac{16}{7}$

14. $\frac{9}{11}$

15. $\frac{8}{29}$

Dividing Fractions

Now that you know about reciprocals, and you know how to multiply fractions, dividing fractions will be a breeze.

DIVIDING FRACTIONS is done the same as MULTIPLYING FRACTIONS
EXCEPT
you use the RECIPROCAL of the divisor.

Remember which number is the divisor?

$$18 \div 6 \text{ (divisor)} \qquad \text{(divisor) } 6\overline{)18} \qquad 18 \text{ divided by } 6 \text{ (divisor)}$$

Try dividing $\dfrac{3}{4} \div \dfrac{6}{8}$.

$$\frac{3}{4} \div \frac{6}{8}$$

$\dfrac{6}{8}$ is the divisor, so you need its reciprocal, $\dfrac{8}{6}$.

$$\frac{3}{4} \times \frac{8}{6}$$

Now multiply.

$$\frac{\overset{1}{\cancel{3}}}{\underset{1}{\cancel{4}}} \times \frac{\overset{2}{\cancel{8}}}{\underset{2}{\cancel{6}}} = \frac{1 \times 2}{1 \times 2} = \frac{2}{2}$$

Canceling helps. (CAUTION: Don't cancel until AFTER you've found the reciprocal.)

$$\frac{2}{2} = 1 \qquad \text{The answer is 1.}$$

To Divide Fractions	Invert the divisor. Multiply.

Example Divide: $\dfrac{3}{4} \div \dfrac{2}{7}$.

$$\frac{3}{4} \times \frac{7}{2} = \frac{3 \times 7}{4 \times 2} = \frac{21}{8}$$

Invert the divisor and multiply.

$$\frac{21}{8} = 2\frac{5}{8}$$

In your final answer, change an improper fraction to a whole number or a mixed number.

Let's stop for a moment.

Have you noticed that we keep telling you to reduce your final answer to lowest terms, or, if it is an improper fraction, to write it as a whole number or a mixed number? Making these types of changes in a final answer is called SIMPLIFYING an answer.

To Simplify an Answer	Change an improper fraction to a whole number or a mixed number. Reduce a fraction to its lowest terms.

Back to dividing fractions . . .
What happens if you have mixed numbers or whole numbers in your division problem? No trouble, just change them to improper fractions.

Example   Divide: $7\frac{1}{2} \div 2\frac{2}{4}$.

$$\frac{15}{2} \div \frac{10}{4}$$

First change the mixed numbers to improper fractions.

$$\frac{15}{2} \times \frac{4}{10}$$

Next invert the divisor and multiply.

$$\frac{\overset{3}{\cancel{15}}}{\underset{1}{\cancel{2}}} \div \frac{\overset{2}{\cancel{4}}}{\underset{2}{\cancel{10}}} = \frac{3 \times 2}{1 \times 2} = \frac{6}{2}$$

Cancel.

$$\frac{6}{2} = 3$$

Simplify.
The answer is 3.

Example   Divide: $8 \div 3\frac{1}{5}$.

$$\frac{8}{1} \div \frac{16}{5}$$

Change the numbers to improper fractions.

$$\frac{\overset{1}{\cancel{8}}}{1} \times \frac{5}{\underset{2}{\cancel{16}}} = \frac{1 \times 5}{1 \times 2} = \frac{5}{2}$$

Invert the divisor and multiply.

$$\frac{5}{2} = 2\frac{1}{2}$$

Simplify.
The answer is $2\frac{1}{2}$.

Example

Divide: $3\frac{2}{3} \div \frac{5}{6}$.

$$\frac{11}{3} \div \frac{5}{6}$$ Change the mixed number to an improper fraction.

$$\frac{11}{\overset{1}{\underset{}{3}}} \times \frac{\overset{2}{6}}{5} = \frac{11 \times 2}{1 \times 5} = \frac{22}{5}$$ Invert the divisor and multiply.

$$\frac{22}{5} = 4\frac{2}{5}$$ Simplify.

The answer is $4\frac{2}{5}$.

Exercises

Divide the fractions in each problem below.

1. $\frac{3}{5} \div \frac{6}{7}$

2. $9 \div \frac{3}{4}$

3. $\frac{2}{3} \div \frac{4}{9}$

4. $\frac{1}{2} \div 50$

5. $\frac{3}{5} \div 1\frac{1}{5}$

6. $\frac{3}{4} \div 15$

7. $\frac{3}{4} \div \frac{1}{2}$

8. $3\frac{1}{4} \div 1\frac{1}{6}$

9. $\frac{5}{8} \div \frac{10}{11}$

10. $10 \div \frac{1}{100}$

11. $\frac{23}{100} \div 23$

12. $3\frac{1}{3} \div 2\frac{1}{2}$

13. $7 \div 4\frac{9}{10}$

14. $\frac{7}{10} \div 3$

15. $\frac{5}{6} \div 2\frac{1}{12}$

16. $8\frac{1}{4} \div 1\frac{1}{2}$

Answers

1. $\frac{7}{10}$ **2.** 12   **3.** $1\frac{1}{2}$ **4.** $\frac{1}{100}$ **5.** $\frac{1}{2}$ **6.** $\frac{1}{20}$

7. $1\frac{1}{2}$ **8.** $2\frac{11}{14}$ **9.** $\frac{11}{16}$ **10.** 1000   **11.** $\frac{1}{100}$ **12.** $1\frac{1}{3}$

13. $1\frac{3}{7}$ **14.** $\frac{7}{30}$ **15.** $\frac{2}{5}$ **16.** $5\frac{1}{2}$

Word Problems

Until now we haven't tried any word problems in this unit. Let's go through a few typical examples.

Example   How much is $\frac{1}{2}$ of $\frac{2}{3}$?

Whenever you see "of" used in this way with a fraction, you MULTIPLY.

$$\frac{1}{\cancel{2}_1} \times \frac{\cancel{2}^1}{3} = \frac{1 \times 1}{1 \times 3} = \frac{1}{3}$$

To see that this answer is correct, look at the circle to the right.

$\frac{2}{3}$ of the circle is in color.

How much is $\frac{1}{2}$ of the colored area? $\frac{1}{3}$ $\frac{1}{2}$ of $\frac{2}{3}$ is $\frac{1}{3}$.

Example   Potter's Medical Supply Company does a substantial amount of overseas business. In order to ensure fast mail delivery at a reasonable price, the company keeps the weights of all its packages below 19 pounds (304 ounces). For a certain package, they want to include 15 items, each weighing $11\frac{1}{3}$ ounces. The packing materials weigh 95 oz. Will the total weight be under 304 ounces?

Multiply 15 by $11\frac{1}{3}$. $15 \times 11\frac{1}{3} = \frac{\cancel{15}^5}{1} \times \frac{34}{\cancel{3}_1} = \frac{170}{1} = 170$

Now add 170 plus 95.

$$\begin{array}{r} 170 \\ +\ 95 \\ \hline 265 \end{array}$$ The package will weigh 265 ounces. It is under 304 ounces.

Example   At an all-day auction, $\frac{5}{6}$ of the total available merchandise was sold. $\frac{3}{4}$ of what was sold was sold in the afternoon. What fraction of the total available merchandise was sold in the afternoon?

In all, $\frac{5}{6}$ of the merchandise was sold. You know that $\frac{3}{4}$ of the merchandise that was sold was sold in the afternoon. Or, $\frac{3}{4}$ of $\frac{5}{6}$ was sold in the afternoon. So what you want to know is: how much is $\frac{3}{4}$ of $\frac{5}{6}$? Whenever you see "of" used in this way with a fraction, you multiply.

$$\frac{\cancel{3}^1}{4} \times \frac{5}{\cancel{6}_2} = \frac{1 \times 5}{4 \times 2} = \frac{5}{8}$$

The auctioneer sold $\frac{5}{8}$ of the total merchandise in the afternoon.

Now study these word problems, which involve dividing fractions.

Example   Joan Thoms helps to figure out the flying times for an airline company. The distance between two cities is 1,980 miles, and the company jet averages about 360 miles per hour. How long will the flight take?

You divide 1,980 by 360.

$$360\overline{)1980}$$
$$\underline{1800}$$
$$180$$

with quotient 5

Write the answer as a mixed number.

$\dfrac{180}{360} = \dfrac{1}{2}$ The flight will take $5\dfrac{1}{2}$ hours.

Example   Jeff Wick, a welder, has a metal strip which he wishes to cut into smaller pieces. Each piece has to be $3\dfrac{1}{3}$ inches long. If the metal strip is $26\dfrac{2}{3}$ inches long, how many pieces can Jeff cut?

You divide $26\dfrac{2}{3}$ by $3\dfrac{1}{3}$. $26\dfrac{2}{3} \div 3\dfrac{1}{3}$

$$\dfrac{80}{3} \div \dfrac{10}{3}$$

$$\dfrac{\overset{8}{\cancel{80}}}{\cancel{3}} \times \dfrac{\cancel{3}}{\cancel{10}} = \dfrac{8 \times 1}{1 \times 1} = \dfrac{8}{1} = 8$$

8 pieces of $3\dfrac{1}{3}$-inch metal can be cut from a $26\dfrac{2}{3}$-inch strip.

Example   Routinely, as newly designed automobiles are produced, a team of automotive technicians tests their performance. Several tests record gas consumption. One technician drove a new car $57\dfrac{1}{2}$ miles and used $3\dfrac{1}{4}$ gallons of gas. How many miles per gallon did the car average?

You need to divide $57\dfrac{1}{2}$ by $3\dfrac{1}{4}$. $57\dfrac{1}{2} \div 3\dfrac{1}{4}$

$$\dfrac{115}{2} \div \dfrac{13}{4}$$

$$\dfrac{115}{\cancel{2}} \div \dfrac{\overset{2}{\cancel{4}}}{13} = \dfrac{115 \times 2}{1 \times 13} = \dfrac{230}{13} = 17\dfrac{9}{13}$$

The car averaged $17\dfrac{9}{13}$ miles per gallon.

Try the word problems on the next page.

Exercises

1. Many cities are required by ordinance to provide a specified amount of park land for their residents. How many acres of parkland are there in a city with 24 parks each averaging $9\frac{1}{3}$ acres?

2. Brian McCandless goes shopping with $54 in his pocket. If he spends $\frac{2}{3}$ of his money, how much money did he spend?

3. What is the cost of $2\frac{5}{8}$ pounds of vegetables at 32¢ per pound?

4. Evelyn Kruger is baking bread. She has $9\frac{1}{3}$ ounces of yeast. If she needs $1\frac{1}{3}$ ounces for each loaf of bread, how many loaves can she bake?

5. Judy Iwata drove her moped $34\frac{1}{4}$ miles in $1\frac{2}{3}$ hours. What was her average speed?

6. According to the encyclopedia, a snake travels about $2\frac{1}{2}$ miles per hour. If it traveled for 10 hours to its yearly breeding grounds, how far would this be?

7. John and Donna build and sell coffee tables in their spare time. They make them by cutting old tree trunks left by highway crews into slices. They found a 12-foot trunk and cut it into 4-inch slices with a chain saw. How many slices did they get?

8. Bernie Everett, manager of a wholesale warehouse, found that his crews were able to load $3\frac{1}{2}$ trucks in only $1\frac{3}{4}$ hours. How many trucks per hour could they load?

Answers

1. $24 \times 9\frac{1}{3} = 224$ acres   **2.** $54   **3.** $2\frac{5}{8} \times 32 = 84$¢

4. $9\frac{1}{3} \div 1\frac{1}{3} = 7$ loaves   **5.** $20\frac{11}{20}$ miles per hour   **6.** 25 miles

7. 36 slices   **8.** 2 trucks per hour

Multiplying and Dividing Proper Fractions

Here's something different!

PROPER FRACTIONS like $\frac{3}{4}$ and $\frac{1}{2}$ act differently from whole numbers when you multiply or divide by them.

When you multiply two whole numbers, the answer is usually a bigger number.

$$5 \times 2 = 10$$

(Remember that when you multiply any number by 1, you get the same number.)

BUT when you multiply a number by a proper fraction, the answer is a SMALLER number.

$$10 \times \frac{1}{2} = \frac{\overset{5}{\cancel{10}}}{1} \times \frac{1}{\underset{1}{\cancel{2}}} = 5$$

Get ready for another surprise.

When you divide a number by a whole number, your answer is smaller than the first number.

$$6 \div 2 = 3$$

BUT when you divide a number by a proper fraction, your answer is LARGER than the first number.

Here's why.

As the number you divide by $\longrightarrow$ $4\overline{)8}^{\,2}$

gets smaller and smaller, $\longrightarrow$ $2\overline{)8}^{\,4}$

the answer gets bigger $1\overline{)8}^{\,8}$

and bigger! $\frac{1}{2}\overline{)\,8}^{\,16}$ $\longleftarrow$ This is always written as $8 \div \frac{1}{2} = \frac{8}{1} \times \frac{2}{1} = \frac{16}{1}$ or 16.

Example   What is $\frac{4}{5}$ of 20?

$$\frac{4}{5} \times 20 = \frac{4}{\underset{1}{\cancel{5}}} \times \frac{\overset{4}{\cancel{20}}}{1} = 16$$

Example If you need to put 20 pounds of chocolates into $\frac{1}{2}$-pound boxes, how many boxes will it take? You know you will need a lot of boxes, so your answer will be big. Divide 20 by $\frac{1}{2}$.

$$20 \div \frac{1}{2} = \frac{20}{1} \times \frac{2}{1} = 40 \text{ boxes}$$

We have seen that proper fractions act differently from whole numbers when you multiply or divide by them. What about mixed numbers? Mixed numbers like $2\frac{1}{2}$ act much like whole numbers when you multiply or divide by them. When you multiply, you get a bigger number. When you divide, you get a smaller number.

Example What happens when you multiply or divide 10 by $2\frac{1}{2}$?

$10 \times 2\frac{1}{2}$ $\qquad\qquad\qquad$ $10 \div 2\frac{1}{2}$

$\frac{10}{1} \times \frac{5}{2} = 25$ $\qquad\qquad$ $\frac{10}{1} \div \frac{5}{2} = \frac{10}{1} \times \frac{2}{5} = 4$

The answer is LARGER than 10. $\qquad$ The answer is SMALLER than 10.

BUT when you multiply or divide by a proper fraction like $\frac{1}{2}$, everything works the opposite way.

> When you multiply a number by a proper fraction, you get a SMALLER number.
>
> When you divide a number by a proper fraction, you get a LARGER number.

This happens only with PROPER FRACTIONS.

Example What happens when you multiply or divide 10 by $\frac{1}{2}$?

$10 \times \frac{1}{2}$ $\qquad\qquad\qquad$ $10 \div \frac{1}{2}$

$\frac{10}{1} \times \frac{1}{2} = 5$ $\qquad\qquad$ $\frac{10}{1} \div \frac{1}{2} = \frac{10}{1} \times \frac{2}{1} = 20$

The answer is SMALLER than 10. $\qquad$ The answer is LARGER than 10.

To Solve Word Problems With Proper Fractions	Read the problem. Decide whether the answer should be larger or smaller than the whole number in the problem. Divide by the proper fraction if you need a larger number, and multiply by the proper fraction if you need a smaller number.

Exercises

Try these word problems using proper fractions. Keep in mind that you multiply for a small answer, and divide for a big one.

1. Heather Brown makes all her own clothes. By doing this, she feels she saves $\frac{2}{3}$ of the store price. If a skirt costs $18 in the store, how much would Heather save? (Do you want an answer bigger or smaller than $18? If bigger, divide by $\frac{2}{3}$. If smaller, multiply by $\frac{2}{3}$.)

2. Heather is a very thoughtful person. Since her younger sister, Bonnie, isn't old enough to make her own clothes, Heather gave her the money she saved in Exercise 1 to buy a jump suit. However, this was $\frac{1}{2}$ of the amount needed. How much was the jump suit? (Do you want a bigger or smaller number than 12?)

3. You were hiking across country. At the end of the first day, you estimate that you covered 24 miles. This was only $\frac{3}{4}$ of the distance you had hoped to walk. How far had you actually planned to walk the first day?

4. Harold had to split the $60,000 winnings on a lottery ticket among 4 other winners. What was his share?

5. $\frac{7}{8}$ of a class of 32 students graduated at the end of the year. How many students graduated?

6. A group leader gave $10 to the kids in the group and said that it was exactly enough for each to have $\frac{1}{2}$ a dollar. How many kids were in the group?

Answers

1. $\frac{2}{3} \times 18 = \$12$ 2. $12 \div \frac{1}{2} = \$24$ 3. $24 \div \frac{3}{4} = 32$ miles

4. $\frac{1}{5} \times 60,000 = \$12,000$ 5. $\frac{7}{8} \times 32 = 28$ students   6. $10 \div \frac{1}{2} = 20$ kids

We've covered quite a bit of material on fractions in this unit. As a review, work the exercises on this page and the next one. If you have trouble with any, check back in the unit and study the examples and explanations.

Review Exercises

Fill in the missing numbers to make equivalent fractions.

1. $\dfrac{7}{8} = \dfrac{}{56}$

2. $\dfrac{18}{9} = \dfrac{}{3}$

3. $\dfrac{21}{42} = \dfrac{7}{}$

4. $\dfrac{2}{3} = \dfrac{16}{}$

5. $\dfrac{5}{9} = \dfrac{}{36}$

6. $\dfrac{7}{11} = \dfrac{21}{}$

7. $\dfrac{20}{45} = \dfrac{}{9}$

8. $\dfrac{32}{8} = \dfrac{}{1}$

Reduce each fraction to its lowest terms.

9. $\dfrac{28}{400}$

10. $\dfrac{57}{78}$

11. $\dfrac{56}{64}$

12. $\dfrac{45}{95}$

13. $\dfrac{114}{176}$

14. $\dfrac{16}{44}$

15. $\dfrac{120}{600}$

16. $\dfrac{24}{96}$

Change each improper fraction to a whole number or a mixed number.

17. $\dfrac{17}{3}$

18. $\dfrac{22}{4}$

19. $\dfrac{20}{5}$

20. $\dfrac{34}{9}$

21. $\dfrac{26}{8}$

22. $\dfrac{49}{7}$

23. $\dfrac{19}{4}$

24. $\dfrac{144}{12}$

Change each mixed number to an improper fraction.

25. $2\dfrac{6}{7}$

26. $4\dfrac{3}{5}$

27. $9\dfrac{7}{8}$

28. $3\dfrac{9}{13}$

29. $1\dfrac{2}{11}$

30. $7\dfrac{6}{7}$

31. $6\dfrac{1}{2}$

32. $10\dfrac{1}{9}$

Multiply. Simplify the answers.

33. $\dfrac{1}{3} \times \dfrac{7}{8}$

34. $2\dfrac{3}{4} \times 5$

35. $5 \times \dfrac{6}{13}$

36. $7\dfrac{9}{10} \times 3\dfrac{1}{2}$

37. $2\dfrac{3}{4} \times 2\dfrac{3}{4}$

38. $3\dfrac{1}{6} \times \dfrac{1}{3}$

39. $4 \times 4\dfrac{7}{8}$

40. $\dfrac{9}{11} \times \dfrac{8}{9}$

Divide. Simplify the answers.

41. $\dfrac{2}{7} \div \dfrac{9}{10}$ **42.** $3\dfrac{7}{8} \div 4$ **43.** $7\dfrac{3}{4} \div \dfrac{2}{3}$ **44.** $6\dfrac{1}{2} \div 3\dfrac{1}{4}$

45. $\dfrac{2}{5} \div 3$ **46.** $3\dfrac{1}{3} \div \dfrac{1}{2}$ **47.** $\dfrac{1}{4} \div \dfrac{1}{8}$ **48.** $7\dfrac{1}{3} \div 8\dfrac{2}{5}$

49. Fred Wang works in a chemical laboratory. For one of his experiments, he measures out 14 milliliters of nitric acid. He then divides the nitric acid into 4 equal portions. What is the amount of each portion?

50. A wine maker hires workers to help him harvest the grapes in his vineyard. If he pays them $2 for each basket they fill, how much should he pay a worker who fills $7\dfrac{1}{2}$ baskets?

51. A supermarket ordered 52 dozen eggs. Egg production was low because the chickens were molting (changing to winter feathers), so the farmer could deliver only $\dfrac{1}{4}$ of the order each day over the next four days. How many eggs did the farmer deliver each day?

52. A computer-controlled automobile welder makes 3 spot welds in $\dfrac{1}{4}$ of a second. How many welds can it make per second?

Answers

49. $3\dfrac{1}{2}$ milliliters   **50.** $15   **51.** 13 dozen eggs   **52.** 12 welds

37. $7\dfrac{6}{16}$ **38.** $1\dfrac{18}{?}$ **39.** $19\dfrac{?}{2}$ **40.** $\dfrac{11}{8}$ **41.** $\dfrac{20}{63}$ **42.** $\dfrac{31}{32}$

43. $11\dfrac{5}{8}$ **44.** 2   **45.** $\dfrac{2}{15}$ **46.** $6\dfrac{2}{3}$ **47.** 2   **48.** $\dfrac{55}{63}$

31. $\dfrac{13}{?}$ **32.** $\dfrac{16}{9}$ **33.** $\dfrac{7}{24}$ **34.** $13\dfrac{3}{4}$ **35.** $2\dfrac{4}{13}$ **36.** $27\dfrac{13}{20}$

25. $\dfrac{20}{7}$ **26.** $\dfrac{23}{5}$ **27.** $\dfrac{79}{8}$ **28.** $\dfrac{48}{13}$ **29.** $\dfrac{13}{11}$ **30.** $\dfrac{55}{7}$

19. 4   **20.** $3\dfrac{7}{9}$ **21.** $3\dfrac{1}{4}$ **22.** 7   **23.** $4\dfrac{3}{4}$ **24.** 12

13. $\dfrac{57}{88}$ **14.** $\dfrac{11}{4}$ **15.** $\dfrac{1}{5}$ **16.** $\dfrac{1}{4}$ **17.** $5\dfrac{2}{3}$ **18.** $5\dfrac{1}{2}$

7. 4   **8.** 4   **9.** $\dfrac{7}{100}$ **10.** $\dfrac{19}{26}$ **11.** $\dfrac{7}{8}$ **12.** $\dfrac{9}{19}$

1. 49   **2.** 6   **3.** 14   **4.** 24   **5.** 20   **6.** 33

Unit 5 Test

Use a separate piece of paper to work out each problem.

Change the mixed numbers to improper fractions, and the improper fractions to mixed numbers.

1. $4\frac{3}{5}$

2. $\frac{11}{5}$

3. $2\frac{2}{3}$

4. $3\frac{2}{5}$

5. $4\frac{5}{7}$

6. $\frac{12}{7}$

Multiply. Simplify the answers.

7. $7\frac{1}{2} \times 2\frac{3}{4}$

8. $3\frac{2}{3} \times \frac{5}{6}$

9. $\frac{4}{5} \times 10$

10. $8 \times 2\frac{1}{4}$

11. $\frac{3}{4} \times \frac{5}{8}$

12. $\frac{2}{7} \times 1\frac{1}{2}$

13. What is the reciprocal of $3\frac{1}{4}$?

Divide. Simplify the answers.

14. $7\frac{1}{2} \div 2\frac{3}{4}$

15. $3\frac{2}{3} \div \frac{5}{6}$

16. $\frac{4}{5} \div 10$

17. $8 \div 2\frac{1}{4}$

18. $\frac{2}{3} \div \frac{4}{7}$

19. $\frac{3}{10} \div 1\frac{1}{3}$

Show all your work for these word problems.

20. George Reath paid $300 in back taxes. However, the tax office informed him that this was only $\frac{4}{5}$ of what he owed. How much were his taxes altogether?

21. Mary Martarono is the payroll clerk for a firm with 642 employees. $\frac{2}{3}$ of the employees are full-time workers. How many employees are full-time workers?

22. A cement company supplies 4 local construction sites with wet mixed cement. If the company has $599\frac{1}{3}$ cubic yards of cement available, how many cubic yards can each site have?

Your Answers

1. _____
2. _____
3. _____
4. _____
5. _____
6. _____
7. _____
8. _____
9. _____
10. _____
11. _____
12. _____
13. _____
14. _____
15. _____
16. _____
17. _____
18. _____
19. _____
20. _____
21. _____
_____
22. _____
_____

6 | Adding and Subtracting Fractions

Aims you toward:

- Knowing how to find the lowest common denominator for a group of fractions.

- Adding fractions and getting the right answer.

- Subtracting fractions and getting the right answer.

In this unit you will work exercises like these:

- Find the lowest common denominator: $\dfrac{5}{7}, \dfrac{1}{6}, \dfrac{2}{3}$.

- Add: $5\dfrac{7}{8} + 3\dfrac{2}{9} + \dfrac{1}{3}$.

- Subtract: $9\dfrac{3}{5} - 7\dfrac{7}{8}$.

Adding Fractions

This circle is divided into fifths. $\frac{2}{5}$ of the circle is shaded in color and $\frac{1}{5}$ of the circle is shaded gray. $\frac{3}{5}$ of the circle is shaded altogether.

$$\frac{2}{5} + \frac{1}{5} = \frac{3}{5}$$

To Add Fractions	If they have the same denominator, add the numerators. The denominator stays the same. Simplify the answer.

Example Add $\frac{7}{9}$ and $\frac{5}{9}$.

$$\frac{7}{9} + \frac{5}{9} = \frac{7+5}{9} = \frac{12}{9}$$

Add the numerators (7 + 5 = 12). The denominator stays the same.

$$\frac{12}{9} = 1\frac{3}{9} = 1\frac{1}{3}$$

Always simplify.
The answer is $1\frac{1}{3}$.

Another Example Add $\frac{2}{7} + \frac{4}{7} + \frac{1}{7}$.

$$\frac{2}{7} + \frac{4}{7} + \frac{1}{7} = \frac{2+4+1}{7} = \frac{7}{7}$$

Add the numerators (2 + 4 + 1 = 7). The denominator stays the same.

$$\frac{7}{7} = 1$$

Simplify.
The answer is 1.

Now try some exercises on adding fractions.

Exercises

1. $\dfrac{4}{5} + \dfrac{3}{5}$

2. $\dfrac{3}{8} + \dfrac{7}{8}$

3. $\dfrac{5}{6} + \dfrac{1}{6}$

4. $\dfrac{2}{3} + \dfrac{2}{3}$

5. $\dfrac{7}{10} + \dfrac{9}{10}$

6. $\dfrac{1}{8} + \dfrac{2}{8}$

7. $\dfrac{7}{9} + \dfrac{8}{9}$

8. $\dfrac{5}{16} + \dfrac{13}{16}$

9. $\dfrac{1}{6} + \dfrac{5}{6} + \dfrac{5}{6}$

10. $\dfrac{3}{12} + \dfrac{5}{12} + \dfrac{11}{12}$

11. $\dfrac{3}{4} + \dfrac{3}{4} + \dfrac{3}{4}$

12. $\dfrac{5}{7} + \dfrac{3}{7} + \dfrac{6}{7}$

13. $\dfrac{5}{7} + \dfrac{2}{7}$

14. $\dfrac{6}{11} + \dfrac{4}{11} + \dfrac{3}{11}$

15. $\dfrac{2}{13} + \dfrac{5}{13}$

16. $\dfrac{1}{2} + \dfrac{1}{2}$

17. Ron Butcher and George Simpson are washing the windows on an apartment building. Ron has finished $\dfrac{1}{6}$ of them, and George has finished $\dfrac{5}{6}$ of them. What fraction of the windows have they done together?

18. Harold Boemester shingled $\dfrac{1}{8}$ of a roof before coffee break, $\dfrac{2}{8}$ after break, another $\dfrac{3}{8}$ after lunch, and $\dfrac{1}{8}$ in the early afternoon. If he finished another $\dfrac{1}{8}$ of the roof after his last break, how much of the roof did he shingle that day?

Answers

17. $\dfrac{6}{6}$ of the windows, or all of them

18. $\dfrac{8}{8}$ of the roof, or all of it

13. 1

14. $1\dfrac{2}{11}$

15. $\dfrac{7}{13}$

16. 1

7. $1\dfrac{2}{3}$

8. $1\dfrac{1}{8}$

9. $1\dfrac{5}{6}$

10. $1\dfrac{7}{12}$

11. $2\dfrac{1}{4}$

12. 2

1. $1\dfrac{2}{5}$

2. $1\dfrac{1}{4}$

3. 1

4. $1\dfrac{1}{3}$

5. $1\dfrac{3}{5}$

6. $\dfrac{3}{8}$

Subtracting Fractions

Suppose you are looking for a snack during a study break. You find $\frac{2}{3}$ of a bag of potato chips in the cabinet. If you eat $\frac{1}{3}$ of what the bag originally held, what fraction of the bag is left? Subtract to find out.

$$\frac{2}{3} \quad - \quad \frac{1}{3} \quad = \quad \frac{1}{3}$$

To Subtract Fractions	If they have the same denominator, subtract the numerators. The denominator stays the same. Simplify the answer.

Example Subtract: $\frac{9}{10} - \frac{3}{10}$.

$$\frac{9}{10} - \frac{3}{10} = \frac{9-3}{10} = \frac{6}{10}$$ Subtract the numerators.
The denominator stays the same.

$$\frac{\overset{3}{\cancel{6}}}{\underset{5}{\cancel{10}}} = \frac{3}{5}$$ Simplify by reducing.
The answer is $\frac{3}{5}$.

Another Example Subtract $\frac{1}{6}$ from $\frac{5}{6}$.

$$\frac{5}{6} - \frac{1}{6} = \frac{5-1}{6} = \frac{4}{6}$$ Subtract the numerators.
The denominator stays the same.

$$\frac{\overset{2}{\cancel{4}}}{\underset{3}{\cancel{6}}} = \frac{2}{3}$$ Simplify.

Exercises on subtracting fractions are on the next page.

Exercises

1. $\dfrac{4}{5} - \dfrac{2}{5}$

2. $\dfrac{5}{7} - \dfrac{2}{7}$

3. $\dfrac{5}{8} - \dfrac{3}{8}$

4. $\dfrac{3}{4} - \dfrac{1}{4}$

5. $\dfrac{8}{9} - \dfrac{2}{9}$

6. $\dfrac{2}{3} - \dfrac{1}{3}$

7. $\dfrac{5}{6} - \dfrac{1}{6}$

8. $\dfrac{9}{10} - \dfrac{3}{10}$

9. $\dfrac{11}{12} - \dfrac{5}{12}$

10. $\dfrac{7}{13} - \dfrac{6}{13}$

11. $\dfrac{5}{9} - \dfrac{4}{9}$

12. $\dfrac{3}{10} - \dfrac{1}{10}$

13. $\dfrac{7}{12} - \dfrac{5}{12}$

14. $\dfrac{4}{5} - \dfrac{1}{5}$

15. $\dfrac{4}{7} - \dfrac{1}{7}$

16. $\dfrac{7}{15} - \dfrac{6}{15}$

17. The measurements of the gears and springs in a wristwatch must be extremely precise. If a gear measures $\dfrac{83}{90}$ of a centimeter in diameter, and it should measure $\dfrac{79}{90}$ of a centimeter, how much must be polished off?

18. A jeweler has $\dfrac{6}{7}$ of an ounce of 14 carat gold. If she uses $\dfrac{4}{7}$ of an ounce to make a ring, how much gold does she have left?

Answers

1. $\dfrac{2}{5}$

2. $\dfrac{3}{7}$

3. $\dfrac{1}{4}$

4. $\dfrac{1}{2}$

5. $\dfrac{2}{3}$

6. $\dfrac{1}{3}$

7. $\dfrac{2}{3}$

8. $\dfrac{3}{5}$

9. $\dfrac{1}{4}$

10. $\dfrac{1}{13}$

11. $\dfrac{1}{9}$

12. $\dfrac{1}{5}$

13. $\dfrac{1}{6}$

14. $\dfrac{3}{5}$

15. $\dfrac{3}{7}$

16. $\dfrac{1}{15}$

17. $\dfrac{2}{45}$ of a centimeter

18. $\dfrac{2}{7}$ of an ounce

Common Denominators

What happens when you want to add two fractions like $\frac{1}{3}$ and $\frac{2}{5}$?

You can't just add the numerators the way you did before, because the denominators are not the same. Fractions like these are called UNLIKE FRACTIONS.

To add such fractions you must first learn about COMMON DENOMINATORS. When fractions have a COMMON DENOMINATOR, it simply means that they have the same denominator.

$\frac{3}{8}$ and $\frac{5}{8}$ are fractions with a common denominator.

They are also called LIKE FRACTIONS.

Often you have fractions that do not have a common denominator and you must write them as fractions that do have a common denominator.

Example   Write $\frac{1}{3}$ and $\frac{2}{5}$ as equivalent fractions with a common denominator.

First change $\frac{1}{3}$ to an equivalent fraction with 15 as the denominator.

How did we get 15? Well, we just think of a number that both 3 and 5 will divide into evenly. (We talked about equivalent fractions on page 78.)

$$\frac{1 \times 5}{3 \times 5} = \frac{5}{15}$$

Now that you know how to multiply fractions, you can write the process like this:

$$\frac{1}{3} \times \frac{5}{5} = \frac{5}{15}.$$

Now change $\frac{2}{5}$ to an equivalent fraction with 15 as the denominator.

$$\frac{2}{5} \times \frac{3}{3} = \frac{6}{15}$$

Take a look at the two new fractions: $\frac{5}{15} = \frac{1}{3}$ and $\frac{6}{15} = \frac{2}{5}$.

$\frac{5}{15}$ and $\frac{6}{15}$ are fractions with a common denominator.

On the next page we'll talk about a method for finding common denominators that are a little harder.

To find the common denominator for two fractions, begin by listing some MULTIPLES of both denominators. Let's continue using $\frac{1}{3}$ and $\frac{2}{5}$ as our example.

Start by listing some multiples of 3. (Find these by multiplying 3×1, 3×2, 3×3, 3×4, 3×5, etc.) Then list some multiples of 5. Obviously you can't list all of the multiples, because the list would be too long. Just list the first few. The first 9 numbers in each list come from the multiplication tables that you memorized in Unit 3.

Here are the multiples of 3, starting with numbers from your 3 times table in Unit 3.

$$3 \quad 6 \quad 9 \quad 12 \quad \textcircled{15} \quad 18 \quad 21 \quad 27 \quad \textcircled{30} \quad 33.$$

Now use your 5 times table: $5 \quad 10 \quad \textcircled{15} \quad 20 \quad 25 \quad \textcircled{30} \quad 35 \quad 40 \quad 45 \quad 50.$

You want a number that is in both of these lists. 15 is in both lists. So is 30. Always choose the lowest number that is in both lists. Here it is 15. The number 15 is called the LOWEST COMMON MULTIPLE of 3 and 5.

Since 3 and 5 are denominators in fractions, and we are looking for a common denominator, 15 is the LOWEST COMMON DENOMINATOR, or LCD, for $\frac{1}{3}$ and $\frac{2}{5}$.

To Find the LCD	List the multiples of each denominator. The LCD is the lowest common multiple.

After you find the lowest common denominator for the fractions, the next step is to write them as equivalent fractions with the LCD as the new denominator.

Example   Write $\frac{1}{2}$, $\frac{1}{3}$, and $\frac{5}{8}$ as fractions with a lowest common denominator.

First find the LCD for the fractions. List some multiples of 2, 3, and 8.

(2 times table) $2 \quad 4 \quad 6 \quad 8 \quad 10 \quad 12 \quad 14 \quad 16 \quad 18 \quad 20 \quad 22 \quad \textcircled{24}$

(3 times table) $3 \quad 6 \quad 9 \quad 12 \quad 15 \quad 18 \quad 21 \quad \textcircled{24} \quad 27 \quad 30$

(8 times table) $8 \quad 16 \quad \textcircled{24} \quad 32 \quad 40$

LCD: 24

Change each fraction to an equivalent fraction with 24 as the denominator.

$$\frac{1}{2} \times \frac{12}{12} = \frac{12}{24} \qquad\qquad \frac{1}{3} \times \frac{8}{8} = \frac{8}{24} \qquad\qquad \frac{5}{8} \times \frac{3}{3} = \frac{15}{24}$$

Exercises

Write each pair of fractions as fractions with a lowest common denominator.

1. $\dfrac{3}{4}$ $\dfrac{1}{3}$　　2. $\dfrac{1}{3}$ $\dfrac{5}{6}$　　3. $\dfrac{5}{8}$ $\dfrac{1}{4}$　　4. $\dfrac{1}{6}$ $\dfrac{7}{12}$

5. $\dfrac{1}{2}$ $\dfrac{1}{3}$　　6. $\dfrac{4}{5}$ $\dfrac{1}{2}$　　7. $\dfrac{2}{3}$ $\dfrac{3}{5}$　　8. $\dfrac{1}{6}$ $\dfrac{3}{8}$

9. $\dfrac{3}{4}$ $\dfrac{2}{5}$　　10. $\dfrac{7}{8}$ $\dfrac{1}{3}$　　11. $\dfrac{4}{5}$ $\dfrac{1}{3}$　　12. $\dfrac{3}{8}$ $\dfrac{7}{12}$

13. $\dfrac{5}{8}$ $\dfrac{3}{16}$　　14. $\dfrac{2}{3}$ $\dfrac{1}{8}$　　15. $\dfrac{3}{4}$ $\dfrac{1}{6}$　　16. $\dfrac{5}{6}$ $\dfrac{7}{9}$

17. $\dfrac{3}{8}$ $\dfrac{7}{12}$　　18. $\dfrac{5}{6}$ $\dfrac{3}{8}$

19. Jack, who was carrying $\dfrac{2}{3}$ of a pail of water, told Jill that he was carrying more than she was. Jill's pail was $\dfrac{7}{10}$ full. Was Jack right?

20. What would you rather have, $\dfrac{4}{5}$ of a pound of silver, or $\dfrac{5}{8}$ of a pound of silver?

Comparing Fractions

Common denominators can be used to compare the size of fractions. Suppose you want to know which is the larger of these two fractions, $\frac{5}{6}$ or $\frac{2}{3}$.

Change $\frac{5}{6}$ and $\frac{2}{3}$ to fractions with a common denominator.

Find the LCD.

Some multiples of 6:   ⑥   12   18

Some multiples of 3:   3   ⑥   9

LCD: 6

$\frac{2}{3} \times \frac{2}{2} = \frac{4}{6}$ Change $\frac{2}{3}$ to an equivalent fraction with 6 as the denominator. $\frac{5}{6}$ does not need to be changed, since the denominator is already 6.

You know that $\frac{5}{6}$ is larger than $\frac{4}{6}$, so $\frac{5}{6}$ is also larger than $\frac{2}{3}$.

You can also write $\frac{5}{6} > \frac{2}{3}$. The $>$ sign means "is greater than." If you flip the sign around, $<$, it means "is less than." $\frac{2}{3} < \frac{5}{6}$

Example Put these three fractions in order, from largest to smallest: $\frac{4}{5}, \frac{2}{3}, \frac{7}{10}$.

Find the LCD.

Some multiples of 5:   5   10   15   20   25   ㉚   35

Some multiples of 3:   3   6   9   12   15   18   21   24   27   ㉚

Some multiples of 10:   10   20   ㉚

LCD: 30

$\frac{4}{5} \times \frac{6}{6} = \frac{24}{30}$ $\frac{2}{3} \times \frac{10}{10} = \frac{20}{30}$ $\frac{7}{10} \times \frac{3}{3} = \frac{21}{30}$

Change the fractions to equivalent fractions with denominators of 30.

You know that $\frac{24}{30} > \frac{21}{30} > \frac{20}{30}$.

$$\downarrow \qquad \downarrow \qquad \downarrow$$

So $\frac{4}{5} > \frac{7}{10} > \frac{2}{3}$.

Exercises

Compare the following pairs of numbers. Use these signs: $>$, $<$, or $=$. Problems 1 and 3 are done for you.

1. $\dfrac{3}{5}$ $\dfrac{7}{8}$

LCD: 40

$\dfrac{3}{5} = \dfrac{24}{40}$ $\dfrac{7}{8} = \dfrac{35}{40}$

$\dfrac{24}{40} < \dfrac{35}{40}$, so $\dfrac{3}{5} < \dfrac{7}{8}$.

2. $\dfrac{3}{8}$ $\dfrac{1}{4}$

3. $3\dfrac{2}{3}$ $3\dfrac{5}{6}$

$3\dfrac{2}{3} = \dfrac{11}{3} \times \dfrac{2}{2} = \dfrac{22}{6}$

$3\dfrac{5}{6} = \dfrac{23}{6} \times \dfrac{1}{1} = \dfrac{23}{6}$

$\dfrac{22}{6} < \dfrac{23}{6}$, so $3\dfrac{2}{3} < 3\dfrac{5}{6}$.

4. $\dfrac{2}{5}$ $\dfrac{7}{10}$

5. $5\dfrac{3}{5}$ $5\dfrac{1}{2}$
6. $\dfrac{3}{4}$ $\dfrac{2}{3}$
7. $8\dfrac{2}{3}$ $8\dfrac{6}{9}$
8. $\dfrac{1}{3}$ $\dfrac{1}{2}$

9. $2\dfrac{3}{4}$ $2\dfrac{5}{6}$
10. $\dfrac{3}{4}$ $\dfrac{5}{6}$
11. $6\dfrac{2}{3}$ $6\dfrac{5}{8}$
12. $\dfrac{1}{3}$ $\dfrac{2}{3}$

13. $\dfrac{2}{3}$ $\dfrac{5}{12}$
14. $\dfrac{1}{2}$ $\dfrac{4}{8}$
15. $6\dfrac{1}{2}$ $6\dfrac{2}{3}$
16. $\dfrac{1}{2}$ $\dfrac{3}{5}$

17. $9\dfrac{1}{3}$ $9\dfrac{1}{4}$
18. $1\dfrac{4}{5}$ $1\dfrac{2}{3}$
19. $\dfrac{3}{5}$ $\dfrac{2}{3}$
20. $\dfrac{1}{3}$ $\dfrac{3}{5}$

21. Would you rather inherit $\dfrac{5}{9}$ or $\dfrac{2}{3}$ of a rich aunt's estate?

22. You and a friend split a bottle of chocolate milk by drinking it together, through two straws. Your friend drank $\dfrac{3}{8}$ of it, and you drank $\dfrac{10}{16}$ of it. Who drank more?

Answers

21. If you're like most of us, you'd rather have $\dfrac{2}{3}$ of the estate. 22. You did.

1. $>$ 2. $>$ 3. $<$ 4. $<$ 5. $>$ 6. $>$
7. $=$ 8. $<$ 9. $<$ 10. $<$ 11. $>$ 12. $<$
13. $>$ 14. $=$ 15. $<$ 16. $<$ 17. $>$ 18. $>$
19. $<$ 20. $<$

Adding Unlike Fractions

Now that you know about common denominators, we can get back to adding unlike fractions. Suppose, for example, that you want to add $\frac{3}{5}$ and $\frac{2}{7}$.

First find the LCD. List multiples of 5 and multiples of 7.

(5 times table)　5　15　20　25　30　㉟　40

(7 times table)　7　14　21　28　㉟　42

LCD:　35

Next make equivalent fractions with the LCD as the denominator.

$$\frac{2}{7} \times \frac{5}{5} = \frac{10}{35} \quad \text{and} \quad \frac{3}{5} \times \frac{7}{7} = \frac{21}{35}$$

Now add.

$$\frac{3}{5} + \frac{2}{7} = \frac{21}{35} + \frac{10}{35} = \frac{31}{35}$$

To Add Unlike Fractions	Find the LCD. Change the fractions to equivalent fractions with the LCD as the denominator. Add. Simplify the answer if possible.

Another Example　Add $\frac{5}{6}$ and $\frac{1}{3}$.

Find the LCD.

Some multiples of 6:　⑥　12　18

Some multiples of 3:　3　⑥　9　12

LCD:　6

$\frac{1}{3} \times \frac{2}{2} = \frac{2}{6}$　　Make equivalent fractions.
　　　　　　　$\frac{5}{6}$ does not need to be changed.

$\frac{5}{6} + \frac{2}{6} = \frac{7}{6}$　　Now add.

$\frac{7}{6} = 1\frac{1}{6}$　　Simplify.

Exercises

1. $\dfrac{3}{4} + \dfrac{3}{8}$ 2. $\dfrac{1}{2} + \dfrac{3}{4}$ 3. $\dfrac{7}{8} + \dfrac{5}{16}$ 4. $\dfrac{1}{2} + \dfrac{5}{6}$

5. $\dfrac{3}{10} + \dfrac{4}{5}$ 6. $\dfrac{7}{10} + \dfrac{1}{2}$ 7. $\dfrac{11}{12} + \dfrac{1}{6}$ 8. $\dfrac{3}{4} + \dfrac{1}{3}$

9. $\dfrac{4}{5} + \dfrac{1}{2}$ 10. $\dfrac{2}{3} + \dfrac{4}{5}$ 11. $\dfrac{3}{4} + \dfrac{3}{5}$ 12. $\dfrac{5}{8} + \dfrac{7}{12}$

13. $\dfrac{3}{7} + \dfrac{6}{7}$ 14. $\dfrac{2}{3} + \dfrac{7}{12}$ 15. $\dfrac{2}{3} + \dfrac{1}{2}$ 16. $\dfrac{1}{2} + \dfrac{4}{5}$

17. In the spring, Bill collected sap from maple trees while his wife, Molly, boiled it into syrup. On one tree he collected $\dfrac{2}{5}$ of a pail in the morning, $\dfrac{1}{2}$ of a pail at noon, and $\dfrac{3}{10}$ of a pail in the evening. How many pails did this tree yield?

18. Mr. Aggie watches his budget carefully. While painting his house, he used $\dfrac{3}{8}$ of a gallon of paint on the window trim and $\dfrac{5}{8}$ of a gallon on the doors. How much paint did he use altogether?

Answers

17. $1\dfrac{1}{5}$ pails   18. $\dfrac{8}{8}$ or 1 gallon

16. $1\dfrac{3}{10}$

11. $1\dfrac{7}{20}$ 12. $1\dfrac{5}{24}$ 13. $1\dfrac{2}{7}$ 14. $1\dfrac{1}{4}$ 15. $1\dfrac{5}{6}$

6. $1\dfrac{1}{5}$ 7. $1\dfrac{1}{12}$ 8. $1\dfrac{1}{12}$ 9. $1\dfrac{3}{10}$ 10. $1\dfrac{7}{15}$

1. $1\dfrac{1}{8}$ 2. $1\dfrac{1}{4}$ 3. $1\dfrac{3}{16}$ 4. $1\dfrac{1}{3}$ 5. $1\dfrac{1}{10}$

Now try the following exercises for some more practice.

Exercises

1. $\dfrac{2}{3} + \dfrac{4}{5}$ **2.** $\dfrac{5}{7} + \dfrac{1}{3}$ **3.** $\dfrac{1}{5} + \dfrac{3}{4} + \dfrac{1}{10}$ **4.** $\dfrac{5}{8} + \dfrac{1}{2} + \dfrac{5}{6}$

5. Last winter, Alice wanted to keep cold air from getting in under her door. She tacked down a strip of wood $\dfrac{1}{4}$ of an inch thick. The cold air still came in, so she added another strip $\dfrac{1}{8}$ inch thick and still another $\dfrac{3}{16}$ inch thick. She needed a thickness of $\dfrac{9}{16}$ inch altogether. Were the three pieces of wood enough?

6. Bill's truck was licensed to carry $\dfrac{3}{4}$ of a ton. He wanted to take 3 calves to market. Normally, calves are weighed in pounds, but Bill used tons in order to keep track of the weight in fractions. The calves weighed $\dfrac{1}{4}$ of a ton, $\dfrac{3}{8}$ ton, and $\dfrac{3}{16}$ ton. Did Bill make it under the weight limit, or was he overloaded?

7. Which would you rather have, $\dfrac{3}{10}$ of an ounce of gold, or $\dfrac{5}{7}$ of an ounce?

8. $\dfrac{1}{4} + \dfrac{3}{8} + \dfrac{7}{16}$ **9.** $\dfrac{3}{4} + \dfrac{1}{6} + \dfrac{1}{3}$ **10.** $\dfrac{2}{5} + \dfrac{1}{4} + \dfrac{1}{2}$

Answers

1. $1\dfrac{7}{15}$ **2.** $1\dfrac{1}{21}$ **3.** $1\dfrac{1}{20}$ **4.** $1\dfrac{23}{24}$ **5.** Yes **6.** No, because his calves weighed $\dfrac{13}{16}$ of a ton, which is above the limit of $\dfrac{3}{4}$ $\left(\dfrac{3}{4} = \dfrac{12}{16}\right)$. **7.** $\dfrac{5}{7}$. **8.** $1\dfrac{1}{16}$ **9.** $1\dfrac{1}{4}$ **10.** $1\dfrac{3}{20}$ because $\dfrac{5}{7} = \dfrac{50}{70}$ and $\dfrac{3}{10} = \dfrac{21}{70}$.

Adding Mixed Numbers

What happens if you want to add mixed numbers, like $4\frac{1}{10}$ and $3\frac{3}{10}$?

$\frac{1}{10} + \frac{3}{10} = \frac{4}{10}$ First add the fractions.

$4 + 3 = 7$ Then add the whole numbers.

$7\frac{4}{10} = 7\frac{2}{5}$ Put them together and simplify.

Another Example   Add $6 + 3\frac{3}{4}$.

You have only one fraction, so it stays the same.

$6 + 3 = 9$ Add the whole numbers.

$9\frac{3}{4}$ Put the whole number and the fraction together.

Another Example   Add $6\frac{2}{3}$ and $1\frac{3}{4}$.

First find the LCD for $\frac{2}{3}$ and $\frac{3}{4}$.

Some multiples of 3:   3   6   9   ⑫   15

Some multiples of 4:   4   8   ⑫   16

LCD: 12

$\frac{2}{3} \times \frac{4}{4} = \frac{8}{12}$ Change each fraction to an equivalent fraction with 12 as the denominator.

$\frac{3}{4} \times \frac{3}{3} = \frac{9}{12}$

$\frac{8}{12} + \frac{9}{12} = \frac{17}{12}$ Now add. First add the fractions.

$\frac{17}{12} = 1\frac{5}{12}$ Simplify the fraction.

$6 + 1 = 7$ Then add the whole numbers.

$7 + 1\frac{5}{12} = 8\frac{5}{12}$ Add $1\frac{5}{12}$ to 7. The answer is $8\frac{5}{12}$.

Another Example

Add $4 + 17\frac{6}{7} + 2\frac{2}{3} + \frac{4}{5}$.

First find the LCD for $\frac{6}{7}$, $\frac{2}{3}$, and $\frac{4}{5}$.

Some multiples of 7:

7	14	21	28	35	42	49	56
63	70	77	84	91	98	(105)	

Some multiples of 3:

3	6	9	12	15	18	21	24	27
30	33	36	39	42	45	48	51	54
57	60	63	66	69	72	75	78	81
84	87	90	93	96	99	102	(105)	

Some multiples of 5:

5	10	15	20	25	30	35
40	45	50	55	60	65	70
75	80	85	90	95	100	(105)

LCD: 105

(We had to go quite a way in our lists this time. Our method for finding the LCD is sometimes cumbersome, but it always works!)

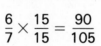

$\frac{6}{7} \times \frac{15}{15} = \frac{90}{105}$

$\frac{2}{3} \times \frac{35}{35} = \frac{70}{105}$ Change the fractions to equivalent fractions with 105 as the denominator.

$\frac{4}{5} \times \frac{21}{21} = \frac{84}{105}$

$\frac{90}{105} + \frac{70}{105} + \frac{84}{105} = \frac{244}{105}$ Now add.

$\frac{244}{105} = 2\frac{34}{105}$ Simplify the fraction.

$4 + 17 + 2 = 23$ Add the whole numbers.

$23 + 2\frac{34}{105} = 25\frac{34}{105}$ Add $2\frac{34}{105}$ to 23.

Exercises

1. $5\frac{1}{3} + 2\frac{2}{3}$ **2.** $4\frac{5}{16} + 1\frac{1}{16}$ **3.** $5 + 1\frac{1}{4}$

4. $6\frac{1}{3} + \frac{1}{3} + 3$ **5.** $2\frac{5}{8} + 1\frac{1}{16} + \frac{3}{4}$ **6.** $4\frac{1}{2} + \frac{5}{6} + 2\frac{2}{3}$

7. $3\frac{7}{10} + \frac{3}{20} + \frac{4}{5}$ **8.** $\frac{1}{8} + 5 + 3\frac{5}{8} + 2\frac{1}{8}$ **9.** $4\frac{1}{2} + \frac{5}{7} + 2\frac{13}{14}$

10. $10\frac{1}{4} + 1\frac{5}{12} + 2\frac{1}{2}$ **11.** $\frac{1}{30} + \frac{4}{5} + 2\frac{5}{6} + \frac{1}{3}$ **12.** $\frac{3}{8} + \frac{1}{2} + \frac{11}{30}$

13. Each week a door-to-door salesman must make a mileage report to his supervisor. On Monday the salesman had to pick up his supplies, so he drove $57\frac{1}{4}$ miles. He drove $12\frac{3}{10}$ miles on Tuesday, $18\frac{1}{2}$ miles on Wednesday, and $29\frac{3}{4}$ miles on Thursday. How many total miles should he write on his mileage report?

14. Vernus McLay's favorite winter pastime is watching hockey. On Monday, she spent $2\frac{1}{2}$ hours at the ice arena. Wednesday night she saw an overtime game that lasted for $3\frac{1}{4}$ hours. She spent $5\frac{1}{3}$ hours at a Saturday tournament, and Sunday Vernus discussed her son's progress with Coach Thompson for $\frac{3}{4}$ hour. How much time did she spend that week watching or discussing hockey?

Answers

1. 8   **2.** $5\frac{3}{8}$ **3.** $6\frac{1}{4}$ **4.** $9\frac{2}{3}$ **5.** $4\frac{7}{16}$

6. 8   **7.** $4\frac{13}{20}$ **8.** $10\frac{7}{8}$ **9.** $8\frac{1}{7}$ **10.** $14\frac{1}{6}$

11. 4   **12.** $1\frac{29}{120}$ **13.** $117\frac{4}{5}$ miles   **14.** $11\frac{5}{6}$ hours

Subtracting Unlike Fractions

Let's take a look now at subtracting unlike fractions. You subtract unlike fractions in the same way you add them.

Example Subtract: $\dfrac{7}{8} - \dfrac{1}{2}$.

First find the LCD.

Some multiples of 8: 8   16   24   32

Some multiples of 2: 2   4   6   8   10

LCD: 8

$\dfrac{1}{2} \times \dfrac{4}{4} = \dfrac{4}{8}$ Make equivalent fractions.
$\dfrac{7}{8}$ does not need to be changed.

$\dfrac{7}{8} - \dfrac{4}{8} = \dfrac{3}{8}$ Now subtract.
The answer is $\dfrac{3}{8}$.

To Subtract Unlike Fractions	Find the LCD. Write the fractions as equivalent fractions with the LCD as the denominator. Subtract. Simplify the answer if possible.

Another Example Subtract $\dfrac{1}{5}$ from $\dfrac{4}{6}$. (This is the same as $\dfrac{4}{6} - \dfrac{1}{5}$.)

First find the LCD.

Some multiples of 6: 6   12   18   24   30   36

Some multiples of 5: 5   10   15   20   25   30   35

LCD: 30

$\dfrac{4}{6} \times \dfrac{5}{5} = \dfrac{20}{30}$ Make equivalent fractions.

$\dfrac{1}{5} \times \dfrac{6}{6} = \dfrac{6}{30}$

$\dfrac{20}{30} - \dfrac{6}{30} = \dfrac{14}{30}$ Now you can subtract.
The answer is $\dfrac{14}{30} = \dfrac{7}{15}$.

Exercises

Subtract to find the answer to each problem below.

1. $\dfrac{5}{6} - \dfrac{2}{3}$ 2. $\dfrac{7}{8} - \dfrac{1}{2}$ 3. $\dfrac{1}{2} - \dfrac{1}{4}$ 4. $\dfrac{5}{8} - \dfrac{1}{2}$

5. $\dfrac{3}{4} - \dfrac{1}{2}$ 6. $\dfrac{6}{7} - \dfrac{2}{7}$ 7. $\dfrac{2}{3} - \dfrac{1}{6}$ 8. $\dfrac{2}{3} - \dfrac{1}{4}$

9. $\dfrac{5}{6} - \dfrac{2}{3}$ 10. $\dfrac{1}{2} - \dfrac{1}{16}$ 11. $\dfrac{4}{5} - \dfrac{3}{10}$ 12. $\dfrac{7}{8} - \dfrac{5}{8}$

13. $\dfrac{1}{7} - \dfrac{2}{14}$ 14. $\dfrac{1}{3} - \dfrac{1}{6}$ 15. $\dfrac{1}{2} - \dfrac{3}{10}$ 16. $\dfrac{5}{6} - \dfrac{3}{4}$

17. $\dfrac{3}{7} - \dfrac{3}{14}$ 18. $\dfrac{6}{11} - \dfrac{1}{2}$ 19. $\dfrac{1}{5} - \dfrac{1}{10}$

20. Martin decided to use a Holland theme in landscaping his front lawn. He bought $\dfrac{5}{8}$ of a ton of topsoil and put $\dfrac{1}{3}$ of a ton around a small windmill. How much topsoil was left for Martin to use in filling two flower boxes shaped like wooden shoes?

Answers

20. $\dfrac{7}{24}$ of a ton of topsoil

16. $\dfrac{1}{12}$ 17. $\dfrac{3}{14}$ 18. $\dfrac{1}{22}$ 19. $\dfrac{1}{10}$

11. $\dfrac{1}{2}$ 12. $\dfrac{1}{4}$ 13. 0   14. $\dfrac{1}{6}$ 15. $\dfrac{1}{5}$

6. $\dfrac{4}{7}$ 7. $\dfrac{1}{2}$ 8. $\dfrac{5}{12}$ 9. $\dfrac{1}{6}$ 10. $\dfrac{7}{16}$

1. $\dfrac{1}{6}$ 2. $\dfrac{3}{8}$ 3. $\dfrac{1}{4}$ 4. $\dfrac{1}{8}$ 5. $\dfrac{1}{4}$

Subtracting Mixed Numbers

If you have a subtraction problem with one or more mixed numbers, change them to improper fractions before you subtract.

Example   Subtract: $4\frac{1}{3} - \frac{2}{3}$.

$4\frac{1}{3} = \frac{13}{3}$ Change the mixed number to an improper fraction.

$\frac{13}{3} - \frac{2}{3}$ Now the problem looks like this.
Subtract as you would any like fractions.

$\frac{13}{3} - \frac{2}{3} = \frac{11}{3}$

$\frac{11}{3} = 3\frac{2}{3}$ Simplify.

Another Example   Subtract: $7\frac{1}{6} - 3\frac{1}{3}$.

$7\frac{1}{6} = \frac{43}{6}$ Change the mixed numbers to improper fractions.

$3\frac{1}{3} = \frac{10}{3}$

$\frac{43}{6} - \frac{10}{3}$ Now the problem looks like this.
Subtract as you would any unlike fractions.

Find the LCD.
Some multiples of 6:   6   12   18
Some multiples of 3:   3   6   9   12
LCD: 6

$\frac{10}{3} \times \frac{2}{2} = \frac{20}{6}$ Make equivalent fractions.
$\frac{43}{6}$ does not need to be changed.

$\frac{43}{6} - \frac{20}{6} = \frac{23}{6}$ Now you can subtract.

$\frac{23}{6} = 3\frac{5}{6}$ Simplify.

In the examples above you changed mixed numbers to improper fractions. In some subtraction problems you will need to change a whole number to an improper fraction. You'll learn more about this in the exercises on the next page.

Exercises

1. $4\dfrac{1}{4} - 3\dfrac{1}{4}$

2. $5\dfrac{1}{2} - \dfrac{3}{4}$

3. $8 - \dfrac{5}{6}$

$\dfrac{8}{1} - \dfrac{5}{6}$

↑

(You can put a
1 under any
whole number.)

$\dfrac{48}{6} - \dfrac{5}{6}$

4. $8 - 2\dfrac{5}{6}$

$\dfrac{8}{1} - \dfrac{17}{6}$

LCD: 6

$\dfrac{48}{6} - \dfrac{17}{6}$

5. $12\dfrac{1}{2} - 4\dfrac{1}{3}$

6. $10\dfrac{1}{10} - \dfrac{3}{5}$

7. $4\dfrac{11}{15} - 1\dfrac{13}{20}$

(Look for a
larger
denominator.)

8. $1\dfrac{1}{9} - \dfrac{2}{3}$

9. $15\dfrac{1}{8} - 2\dfrac{23}{24}$

10. $7\dfrac{1}{2} - 2$

11. $7 - 2\dfrac{1}{2}$

12. $1\dfrac{5}{6} - \dfrac{6}{7}$

13. Stock in R. S. Steel Company opened one day at $43\dfrac{5}{8}$ on the New York Stock Exchange and closed the same day at $41\dfrac{11}{16}$. What was its loss in price?

14. Al's mother brought home 12 quarts of milk to last all week. By Wednesday Al had finished off $7\dfrac{3}{4}$ quarts. How much was left for the rest of the week?

Answers

1. 1

2. $4\dfrac{3}{4}$

3. $7\dfrac{1}{6}$

4. $5\dfrac{1}{6}$

5. $8\dfrac{1}{6}$

6. $9\dfrac{1}{2}$

7. $3\dfrac{1}{12}$

8. $\dfrac{4}{9}$

9. $12\dfrac{1}{6}$

10. $5\dfrac{1}{2}$

11. $4\dfrac{1}{2}$

12. $\dfrac{41}{42}$

13. $1\dfrac{15}{16}$

14. $4\dfrac{1}{4}$ quarts

Word Problems

The following problems will review the skills you learned in Units 5 and 6.

Exercises

1. Jack and Jill went up the hill.

   It took them $1\frac{3}{4}$ minutes.

   They both fell to the ground.
   And rolled all the way down,

   In $\frac{9}{10}$ the time to climb it.

   How long did it take
   The two of them to make
   The trip from the top to the bottom?

2. If you recall the nursery rhyme, Jack and Jill were supposed to get a pail of water. Actually, they only got $\frac{2}{3}$ of a pail before they fell. On their way down, they spilled $\frac{16}{24}$ of the $\frac{2}{3}$. How much water did they have left by the time they got to the bottom of the hill?

3. On their second attempt, Jack and Jill climbed $\frac{5}{6}$ of the way up, but then stopped to rest. While resting, they figured they had climbed 30 yards. How high was the hill altogether?

4. Jack and Jill had to get the water three times in a row. They were trying to put out a fire they had started on their raft, which was floating in the creek at the bottom of the hill. They carried $4\frac{3}{8}$ pails, then another $6\frac{1}{3}$ pails, and finally $7\frac{11}{12}$ pails. How many pails of water did they carry altogether?

5. Who ever heard of finding water at the TOP of a hill anyway?

Answers

1. $\frac{9}{10}$ of $1\frac{3}{4}$

   $\frac{9}{10} \times \frac{7}{4} = \frac{63}{40}$

   $1\frac{23}{40}$ minutes

2. **Amount spilled:**

   $\overset{2}{\underset{3}{\cancel{\frac{16}{24}}}} \times \frac{2}{3} = \frac{4}{9}$

   Amount left:

   $\frac{2}{3} - \frac{4}{9} = \frac{6}{9} - \frac{4}{9} = \frac{2}{9}$ pail

3. $30 \div \frac{5}{6}$

   $\frac{30}{1} \times \frac{6}{5} = 36$

   36 yards

4. $18\frac{5}{8}$ pails

5. ?

Unit 6 Test

Use a separate piece of paper to work out each problem.

1. $\frac{2}{6} + \frac{3}{4}$

2. $\frac{1}{5} + \frac{3}{8} + \frac{1}{2}$

3. $3\frac{2}{5} + 1\frac{5}{6} + \frac{3}{10}$

4. $\frac{7}{8} + 8\frac{1}{6} + \frac{3}{4}$

5. $21 + \frac{11}{12} + 2\frac{1}{2}$

6. $10\frac{1}{4} + \frac{5}{7} + \frac{3}{14}$

7. $13\frac{1}{15} + 2\frac{1}{12} + \frac{7}{10}$

8. $5\frac{1}{6} + 4\frac{7}{8}$

9. $\frac{5}{6} + \frac{3}{8}$

10. $\frac{7}{8} - \frac{3}{8}$

11. $102\frac{1}{2} - 5\frac{7}{8}$

12. $12 - \frac{3}{4}$

13. $16\frac{1}{3} - 4\frac{5}{6}$

14. $24\frac{1}{8} - 10$

15. $8 - 6\frac{7}{10}$

16. $6\frac{5}{8} - 4\frac{2}{3}$

17. $9\frac{1}{12} - 6\frac{7}{18}$

18. $8\frac{1}{15} - 2\frac{5}{21}$

19. Suppose you were a student at I.V. Tech, and had decided to visit a friend in Santa Barbara for Christmas. You drove $5\frac{3}{4}$ hours on Monday, $11\frac{1}{3}$ hours on Tuesday, $6\frac{1}{8}$ hours on Wednesday, and $9\frac{1}{12}$ hours on Thursday. How many hours did the trip take?

20. Put these fractions in order from largest to smallest: $\frac{7}{8}, \frac{13}{16}, \frac{29}{32}, \frac{3}{4}$. (*Hint:* Find the LCD first.)

21. After swimming $\frac{9}{10}$ of the way across Lake Erie a swimmer got too tired to finish. So he turned around and swam back to the starting place. What fraction of the lake did he swim?

Your Answers

1. _____

2. _____

3. _____

4. _____

5. _____

6. _____

7. _____

8. _____

9. _____

10. _____

11. _____

12. _____

13. _____

14. _____

15. _____

16. _____

17. _____

18. _____

19. _____

20. _____

_____

21. _____

_____

Review Test

Are you having trouble keeping all the operations with fractions in mind? This review test covers multiplication, division, addition, and subtraction of fractions.

1. $\dfrac{2}{7} \times \dfrac{1}{3}$

2. $\dfrac{1}{8} \div \dfrac{3}{5}$

3. $\dfrac{7}{12} - \dfrac{1}{6}$

4. $\dfrac{3}{8} + \dfrac{3}{16}$

5. $2\dfrac{5}{8} - 1\dfrac{3}{4}$

6. $2\dfrac{5}{8} \times 1\dfrac{3}{4}$

7. $2\dfrac{5}{8} \div 1\dfrac{3}{4}$

8. $2\dfrac{5}{8} + 1\dfrac{3}{4}$

9. $6\dfrac{1}{12} \times \dfrac{3}{5}$

10. $13\dfrac{6}{11} - 6\dfrac{21}{22}$

11. $17\dfrac{2}{3} + 3\dfrac{17}{18} + 9\dfrac{1}{2}$

12. $7\dfrac{3}{7} \div 3\dfrac{1}{2}$

13. $4\dfrac{6}{7} - \dfrac{1}{3}$

14. $\dfrac{3}{7} \div \dfrac{1}{2}$

15. $6\dfrac{10}{11} + 9\dfrac{1}{3} + \dfrac{2}{3}$

16. $6\dfrac{2}{3} \times \dfrac{1}{7}$

17. $4\dfrac{1}{4}$ pounds of raw ground meat weighed only $3\dfrac{7}{8}$ pounds after it was cooked. What weight was lost?

18. The owner of a grocery store stocked $\dfrac{1}{4}$ of the store on Tuesday and $\dfrac{3}{8}$ of it on Thursday. What fraction of the store did he stock altogether?

19. A volunteer is stuffing envelopes for a political campaign. He finished $\dfrac{3}{5}$ of them before dinner. He then does $\dfrac{2}{3}$ of the remainder in the evening. What fraction of the total did he do in the evening? (Hint: $\dfrac{2}{5}$ of the total is left after dinner. What is $\dfrac{2}{3}$ of $\dfrac{2}{5}$?)

20. Nancy Oleson works $37\dfrac{1}{2}$ hours a week. If she works 5 days a week, how many hours does she work each day?

Your Answers

1. _____
2. _____
3. _____
4. _____
5. _____
6. _____
7. _____
8. _____
9. _____
10. _____
11. _____
12. _____
13. _____
14. _____
15. _____
16. _____
17. _____
_____
18. _____
_____
19. _____
20. _____
_____

7 | Adding and Subtracting Decimals

Aims you toward:

- Knowing place value for decimals.
- Knowing how to read and write decimals.
- Knowing how to change decimals to fractions.
- Adding decimals and getting the right answer.
- Subtracting decimals and getting the right answer.

In this unit you will work exercises like these:

- Write out the number 89.567.
- Change 0.32 to a fraction.
- Add 78.04 + 45.521 + 3.6
- Subtract: 7.43 − 3.21.

Decimal Fractions

In the last unit you saw that fractions are a way to represent things that are less than one. You can also use DECIMAL FRACTIONS to represent things that are less than one.

A DECIMAL FRACTION is a fraction whose denominator is 10 or 100 or 1000 or some other multiple of 10. Simple enough, isn't it? Let's take a look at an example.

Example $\frac{7}{10}$ is a fraction whose denominator is 10.

In DECIMAL FORM $\frac{7}{10}$ is written 0.7 and is read "seven tenths."

$$\frac{7}{10} = 0.7$$

The symbol "." is called a DECIMAL POINT.

Another Example $\frac{7}{100}$ is a fraction whose denominator is 100. In decimal form $\frac{7}{100}$ is written 0.07 and is read "seven hundredths."

$$\frac{7}{100} = 0.07$$

From now on we will refer to numbers that have a decimal point in them as DECIMALS.

Do you remember the discussion of naming numbers and place value of numbers from way back in Unit 1? Flip back to pages 5–10 if you don't.

Now look at the place values of the decimal shown on the next page.

Naming Decimals

5,183,725.498673

millions
hundred-thousands
ten-thousands
thousands
hundreds
tens
ones
← decimal point
tenths
hundredths
thousandths
ten-thousandths
hundred-thousandths
millionths

The names of the places to the left of the decimal point end in "s." The names of the places to the right of the decimal point end in "ths."

The place values to the left of the decimal point represent numbers that are greater than or equal to one. The place values to the right of the decimal point represent numbers that are less than one. Think of money: $1.99. The "1" on the left of the decimal point is 1 dollar and is greater in value than the ".99" on the right, which is only 99 cents.

To Name Decimals	If there is a number to the left of the decimal point, read it as you would any whole number; then say "and" for the decimal point. Read the number to the right of the decimal point as you would any whole number. Finally, read the name of the place where the last digit falls.

Example   Read the number 5,750.44

"Five thousand seven hundred fifty and forty-four hundredths."

Read the number to the left of the decimal point, then read "and" for the decimal point. Next, read the number to the right of the decimal point. Last, read the name of the place where the last digit falls.

Another Example   How would you read a decimal like 0.603?

"Six hundred three thousandths."

There's nothing to the left of the decimal point, so begin with the number to the right of the decimal point. Read that number as you normally would. Then read the name of the place where the 3 falls.

Note that the only time you say "and" when naming a decimal is for the decimal point.

If a decimal has no number to the left of the decimal point, put a zero in the ones place. The zero does not change the value of the decimal or the way to read it. For example, .79 is usually written as 0.79. Both are correct, and both are read "seventy-nine hundredths."

You can also add zeros to the end of a decimal without changing its value. If you have a number like 3.236, it doesn't change in value if you add zeros to the end.

$$3.236 = 3.2360 = 3.23600 = 3.23600000000$$

Each of these numbers means that you have 3 ones, 2 tenths, 3 hundredths, and 6 thousandths. The added zeros simply tell you that there are no ten-thousandths, no hundred-thousandths, and so on. The zeros don't change the value of the number, but they do change the way the number is read.

3.236 is read "three and two hundred thirty-six thousandths."

3.2360 is read "three and two thousand three hundred sixty ten-thousandths."

3.23600 is read "three and twenty-three thousand six hundred hundred-thousandths."

3.23600000000 is too long to write out!

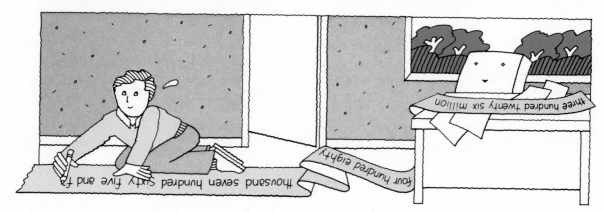

Examples   3,765.4032 is read "three thousand seven hundred sixty-five and four thousand thirty-two ten-thousandths."

390.800 is read "three hundred ninety and eight hundred thousandths."

0.9462 is read "nine thousand four hundred sixty-two ten-thousandths."

0.560 is read "five hundred sixty thousandths."

200.3 is read "two hundred and three tenths."

62.0008 is read "sixty-two and eight ten-thousandths."

A car company may have an engine part that is exactly 4 inches in diameter. On the blueprints, this is shown as 4.000. It's still 4 inches, but this shows that the part is measured to an accuracy of $\frac{1}{1000}$ of an inch.

Exercises

For the decimal 37,196.4852, give the number that has the place value named below. The first two are done to help you get started.

1. Tens = 9   **2.** Tenths = 4   **3.** Hundreds   **4.** Thousandths

5. Ones   **6.** Hundredths   **7.** Thousands   **8.** Ten-thousandths

For the decimal 16,973.8240, give the place value of each number below.

9. 8   **10.** 3   **11.** 2   **12.** 7   **13.** 0

Name each of the following decimals.

14. 7.138   **15.** 13.67   **16.** 310.6   **17.** 2.38

18. 90.0076   **19.** 710.03   **20.** 0.29   **21.** 16.2110

Write each of the following in decimal form.

22. Seven thousand three hundred and four hundredths

23. Sixty-two and nine thousand forty-four ten-thousandths

24. Seven hundred ten thousandths

25. One and one tenth

Answers

1. 9   **2.** 4   **3.** 1   **4.** 5   **5.** 6   **6.** 8
7. 7   **8.** 2   **9.** Tenths   **10.** Ones   **11.** Hundredths
12. Tens   **13.** Ten-thousandths
14. Seven and one hundred thirty-eight thousandths
15. Thirteen and sixty-seven hundredths   **16.** Three hundred ten and six tenths
17. Two and thirty-eight hundredths
18. Ninety and seventy-six ten-thousandths
19. Seven hundred ten and three hundredths   **20.** Twenty-nine hundredths
21. Sixteen and two thousand one hundred ten ten-thousandths   **22.** 7,300.04
23. 62.9044   **24.** 0.710   **25.** 1.1

Comparing Decimal Sizes

Since we've all had practice handling money, let's use money to learn how to compare the sizes of decimal numbers.

Example   Which is bigger, 1.09 or 1.1?

First change the numbers into money.

1.09 becomes $1.09.

1.1   becomes $1.10.

Since 10 is bigger than 9, 1.10 is bigger than 1.09,

or 1.1 > 1.09.

Sometimes a number has too many digits after the decimal point to be compared with money.

Example   Which is bigger, 2 or 0.998?

The number 2 can be written as 2.000. This gives the same number of digits after the decimal point as there are in the number 0.998.

If we pretend that the decimal points are not there, we have 2,000 and 998.

2,000 > 998,
so   2   > 0.998.

Exercises

Try these. The first one is done to get you started.
In each of the following problems, tell which number is bigger, using >.

1. 0.68, 0.8   **2.** 1.2, 1.19   **3.** 0.42, 0.098

0.8 becomes 0.80.
Since 80 is
bigger than 68,
0.8 > 0.68.

4. 4, 0.87   **5.** 0.005, 0.01   **6.** 3.052, 3.502

7. 1, 0.7869   **8.** 7.1, 7.100

Answers

1. 0.8 > 0.68   **2.** 1.2 > 1.19   **3.** 0.42 > 0.098   **4.** 4 > 0.87
5. 0.01 > 0.005   **6.** 3.502 > 3.052   **7.** 1 > 0.7869   **8.** Equal

Changing Decimals to Fractions

There are two methods you can use to change a decimal to a fraction. With the first method, you begin by naming the decimal. Then you simply write the fraction you have named.

Example   Change 0.3 to a fraction.

0.3 is read "three tenths," which is $\frac{3}{10}$.

So, $0.3 = \frac{3}{10}$.

With the other method of changing a decimal to a fraction, you write the number to the right of the decimal point as the numerator. The denominator comes from the place value of the right-hand digit.

Example   Write 0.49 as a fraction.

$0.49 = \frac{49}{100}$ 49 is the numerator.
9 is in the hundredths place, so the denominator is 100.

You can also think of it this way: the denominator has one zero for each place after the decimal point.

$$0.49 = \frac{49}{100}$$

Two places after the decimal point.   Two zeros in the denominator

To Change a Decimal to a Fraction	Name the decimal. Write it as a fraction. Or write the number to the right of the decimal point as the numerator. Use the place value of the last digit to find the denominator. Reduce if needed.

If the decimal is greater than 1, such as 3.85, you change only the part to the right of the decimal point.

Example   Change 3.85 to a fraction.

The 3 is a whole number, so leave it alone.

Change .85 to a fraction.

$0.85 = \frac{85}{100}$ This fraction can be reduced: $\frac{85 \div 5}{100 \div 5} = \frac{17}{20}$.

Add the 3 to the fraction.

$$3.85 = 3\frac{17}{20}$$

Exercises

Change each of the following decimals to fractions. Simplify your answers.

1. 0.07   **2.** 0.001   **3.** 0.013   **4.** 0.33   **5.** 0.4937

6. 4.2   **7.** 10.025   **8.** 0.03   **9.** 0.3   **10.** 0.019

11. 3.16   **12.** 13.06   **13.** 0.98   **14.** 1.6   **15.** 14.08

16. 2.08   **17.** 3.980   **18.** 48.971   **19.** 80.517   **20.** 16.8

21. 47.2   **22.** 16.803   **23.** 36.4   **24.** 26.04

25. If you had your choice between a gift of 0.04 ounces of gold and 0.004 ounces, which would you choose? Why?

26. Bill and Fred argued about a measurement on a blueprint. Bill said that $\frac{78}{100}$ was larger than $\frac{780}{1000}$, and Fred said that $\frac{780}{1000}$ was larger. Who was right?

Answers

26. They were both wrong, because $\frac{78}{100} = \frac{780}{1000}$.

25. 0.04, because it's worth more.

21. $47\frac{1}{5}$ **22.** $16\frac{803}{1000}$ **23.** $36\frac{2}{5}$ **24.** $26\frac{1}{25}$

16. $2\frac{2}{25}$ **17.** $3\frac{49}{50}$ **18.** $48\frac{971}{1000}$ **19.** $80\frac{517}{1000}$ **20.** $16\frac{4}{5}$

11. $3\frac{4}{25}$ **12.** $13\frac{3}{50}$ **13.** $\frac{49}{50}$ **14.** $1\frac{3}{5}$ **15.** $14\frac{2}{25}$

6. $4\frac{1}{5}$ **7.** $10\frac{1}{40}$ **8.** $\frac{3}{100}$ **9.** $\frac{3}{10}$ **10.** $\frac{19}{1000}$

1. $\frac{7}{100}$ **2.** $\frac{1}{1000}$ **3.** $\frac{13}{1000}$ **4.** $\frac{33}{100}$ **5.** $\frac{4937}{10000}$

Adding Decimals

The trick to adding decimals is to keep the digits of the same place value lined up under each other. The easiest way to do this is to LINE UP ALL THE DECIMAL POINTS.

Example   Add 7.6 + 12.8 + 9.23.

$$
\begin{array}{r}
7.6 \\
12.8 \\
+\ \ 9.23 \\
\hline
\end{array}
$$
Line up digits of the same place value by lining up the decimal points.

$$
\begin{array}{r}
\overset{1\ 1}{7.60} \\
12.80 \\
+\ \ 9.23 \\
\hline
29.63
\end{array}
$$
Add.
It may help you add if you put zeros in the open spaces. (Remember that this does not change the values of the numbers.)
Note that you can carry just as you did with whole numbers. Put a decimal point in the answer directly under the other decimal points.

To Add Decimals	Write the numbers under each other with the decimal points lined up. Add the numbers as you normally would. The decimal point in the answer goes directly below the other decimal points.

What happens if you want to add a whole number and a decimal? You can always write a whole number as a decimal without changing its value. To write a whole number as a decimal, just add a decimal point to the right of the number.

Examples   1,940 = 1,940.

45 = 45.

Think of how you use money. You know that $45 = $45.00. The decimal point doesn't change anything.

We don't normally include the decimal point in whole numbers, but it helps when we add whole numbers and decimals.

Example   Add 69 + 8.45 + 23 + 344.9.

$$
\begin{array}{r}
\overset{1\ 2\ 1}{69.00} \\
8.45 \\
23.00 \\
+\ 344.90 \\
\hline
445.35
\end{array}
$$
To make sure you have the columns lined up correctly, add decimal points to the whole numbers.
Add zeros if it helps you.
Add.
Put a decimal point in your answer.

Now try the exercises on adding decimals.

Exercises

1. 10.8 + 7.9	**2.** 0.5 0.204 +0.16	**3.** 0.675 0.9 +0.38	**4.** 4.251 7.9 +0.38

5. 12.48 + 0.673 + 9.1   **6.** 50.001 + 3.96 + 27.8   **7.** 0.09 + 3.6

8. 2.8 + 0.139   **9.** 0.437 + 2.59   **10.** 32.5 + 0.43

11. 8.45 + 0.652   **12.** 0.91 + 0.2 + 0.078   **13.** 0.942 + 0.7 + 0.48

14. 1.2 + 2.86 + 0.093   **15.** 8.725 + 0.23 + 2.46

16. Add: $2.98 + 5 nickels + 15 dollars + 12 dimes + 1 penny.

17. Add: seventy-two hundredths + three and nine hundred twenty-one thousandths + seven and five tenths.

18. The supply manager for Northridge Office Supplies keeps detailed accounts of all sales. The company sold 1.25 gross of pencils on December 8. It sold 0.75 gross on December 10, and 1.50 gross on December 13. How many gross were sold altogether on these three days? (A gross is 144 items.)

19. Five students who were learning the goldsmith trade were given samples of gold dust to weigh on the balance scales. The weights of the samples were as follows: 0.004 ounces, 1 ounce, 0.58 ounces, 1.6 ounces, and 0.19 ounces. How much did all the samples weigh together?

20. Sharon, Sandra, and Heddie spent most of their money shopping. They had to come up with a total of $1.50 to take a bus home. Heddie managed to scrape up 8 cents. Sharon found 3 dimes and 4 nickels, and Sandra came up with 3 quarters, 1 dime, 1 nickel, and 3 pennies. Did they have enough money for bus fare?

Answers

20. Yes, they had $1.51.

16. $19.44   **17.** 12.141   **18.** 3.50 gross   **19.** 3.374 ounces

11. 9.102   **12.** 1.188   **13.** 2.122   **14.** 4.153   **15.** 11.415

6. 81.761   **7.** 3.69   **8.** 2.939   **9.** 3.027   **10.** 32.93

1. 18.7   **2.** 0.864   **3.** 1.955   **4.** 12.531   **5.** 22.253

Subtracting Decimals

You subtract decimals by the same methods used to add decimals.

To Subtract Decimals	Line up the digits of the same place value by lining up the decimal points. Subtract. Put the decimal point in the answer directly below the other decimal points.

Example Subtract: $4.723 - 2.65$.

$$\begin{array}{r} 4.723 \\ -2.65 \\ \hline \end{array}$$ Line up the numbers with the decimal points under each other.

$$\begin{array}{r} \overset{6}{} \\ 4.\not{7}^1 2\ 3 \\ -2.6\ 5\ 0 \\ \hline 2.0\ 7\ 3 \end{array}$$ Subtract.
Note that you can borrow as you normally would.
Remember to put a decimal point in the correct place.

Another Example Subtract 4.362 from 17.6.

$$\begin{array}{r} 17.6 \\ -\ 4.362 \\ \hline \end{array}$$ Line up the decimal points.
There are no hundredths or thousandths to subtract from, so add zeros.

$$\begin{array}{r} {}^5\ {}^9 \\ 1\ 7.\not{6}\,\not{0}\,{}^1 0 \\ -\ \ 4.3\ 6\ 2 \\ \hline 1\ 3.2\ 3\ 8 \end{array}$$ Now you can subtract.
Remember the decimal point!

Another Example Subtract: $19 - 7.23$.

$$\begin{array}{r} 19. \\ -\ 7.23 \\ \hline \end{array}$$ Put a decimal point after the 19.
Line up the decimal points.

$$\begin{array}{r} {}^8\ {}^9 \\ 1\ \not{9}.\not{0}\,{}^1 0 \\ -\ \ 7.\ 2\ 3 \\ \hline 1\ 1.\ 7\ 7 \end{array}$$ Add zeros.
Subtract.

Try the exercises on the next page.

Exercises

1. 5.9
      −1.86

2. 7.1
      −3.45

3. 4.3
      −2.68

4. 0.59
      −0.372

5. 0.581 − 0.39

6. 7.251 − 2.37

7. 8.75 − 2.6

8. 7.2 − 3

9. 9.4 − 7.25

10. 8.256 − 4.13

11. 4.28 − 2.195

12. 8.104 − 5

13. 8.16 − 3.894

14. 7.639 − 4.1

15. 6.8 − 4.521

16. 7.5 − 3.276

17. 17.365 − 12.19

18. 39.32 − 24.732

19. 49.2 − 36.895

20. 0.9062 − 0.001

21. Laura Nesbit worked 40 hours during a certain week. She worked 8.25 hours on Monday, 6.5 hours on Tuesday, 7.75 hours on Wednesday, and 8.75 hours on Thursday. How many hours did she work on Friday?

22. Wilma lives a long way from a good mechanic, so she decided to service her own car by following the manufacturer's directions exactly. Before servicing, the car went from 0 to 60 miles per hour in 9.2 seconds. Afterwards, it took only 6.08 seconds. How many seconds faster was the car?

Answers

22. 3.12 seconds

21. 8.75 hours

20. 0.9062
19. 12.305
18. 14.588
17. 5.175
16. 4.224
15. 2.279
14. 3.539
13. 4.266
12. 3.104
11. 2.085
10. 4.126
9. 2.15
8. 4.2
7. 6.15
6. 4.881
5. 0.191
4. 0.218
3. 1.62
2. 3.65
1. 4.04

Money and Decimals

Any time you go to a store, or a movie theater, or a bank, or to any place where you use money, you are using decimals. For example,

$83.75 means 83 dollars and 75 cents.

Let's take a look at this as a decimal.

83.75

83 is a whole number.
There are 8 tens and 3 ones, or
8 ten-dollar bills and 3 one-dollar bills.

.75 is a decimal.
There are 7 tenths and 5 hundredths, or
7 dimes and 5 pennies.
A dime is a tenth of a dollar.
A penny is a hundredth of a dollar.

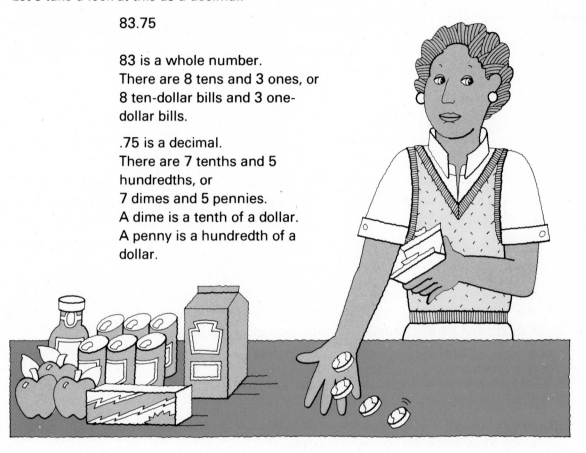

You can treat any expression of money as a decimal. The rules you learned about decimals apply to money as well.

When you add or subtract money, always subtract the dollars from the dollars and the cents from the cents. To do this, line up the decimal points as you would with any decimal.

Example   Add $178.90 + $5.89 + $4 + $3.50.

$$\begin{array}{r} \$179.90 \\ 5.89 \\ 4.00 \\ +\ \ \ 3.50 \\ \hline \$193.29 \end{array}$$
   Line up the decimals ($4 = $4.00)
   Add.

Exercises involving money are on the next page.

Exercises

Solve each of the following problems. Watch out for the signs! ⊕ ⊖

1.	$ 1.10	2.	$13.04	3.	$8.72	4.	$5.43
	78.16		− 6.39		+ 2.46		+ 1.01
	+ 3.30						

5. Find the total value of the following: 6 dollars and 3 cents + 17 dollars and 70 cents + 12 dollars.

6. Subtract 18 cents from 18 dollars.

7. The manager of a local fast-food restaurant keeps records of cash intake. The daily receipts for one week at the restaurant were $1,768.12, $1,412.46, $1,089.73, $1,586.63, $1,986.40, $1,821.43, and $1,470.38. Find the total receipts for the week.

8. Brad Jeffery had a balance of $219.91 in his checking account at the beginning of the month. During the month he made the following deposits: $62.50, $115.90, $75.26, and $400. During the month he also wrote the following checks: $325, $13.63, $29.60, $67.21, $155.77, and $9.25. What was his balance at the end of the month?

9. Dave Vitek spent $67.14 on books, $19.36 on a new pair of Levi's, and $3.72 on lunch. How much did he spend altogether?

10. Mitch, Harold's private nurse, bought 3 bottles of rubbing alcohol for Harold's sore back. The total cost was $8.00. Mitch was able to return one of the bottles unused for a refund of $2.67. How much did Harold have to spend on rubbing alcohol?

Answers

1. $82.56	2. $6.65	3. $11.18	4. $6.44
5. $35.73	6. $17.82	7. $11,135.15	8. $273.11
9. $90.22	10. $5.33		

Word Problems

1. We measure home insulation by using the letter "R." For example, 6 inches of rock wool insulation in a wall is rated as 19 R. Builders tell us never to press or flatten insulation and give as an example the 6 inches mentioned above being compressed to fit into a thinner wall, and then having only a 13.1 R. How many R's of insulation were lost this way?

2. A cross section of an ordinary house wall shows the insulating value of the various materials. How many R's of insulation does this wall have?

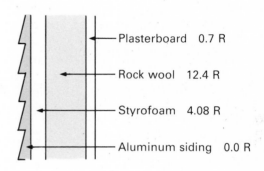

Plasterboard   0.7 R

Rock wool   12.4 R

Styrofoam   4.08 R

Aluminum siding   0.0 R

3. Petty cash is money that most offices have on hand to buy little things needed every day, and no records have to be kept for auditors. In one office, the petty cash spent in one month was as follows: envelopes, $13.47; postage stamps, $24.68; pencils, $8.04; typewriter eraser, 69¢; cellulose tape, $3.98; coffee fund, $31.06; and 7¢ to round off a bank deposit. What was the total amount taken from petty cash?

4. The office manager for the business in Problem 3 was upset about the amount of money taken from petty cash. He had set a limit of $70, and he said that the staff had overspent that amount. Was he right? If so, by how much did the staff go over the limit?

Answers

4. Yes, the staff spent $11.99 over the limit.

1. 5.9 R's   2. 17.18 R's   3. $81.99

Unit 7 Test

Use a separate piece of paper to work out each problem.

Give the place value for the 4 in each number.

1. 1,452.6   **2.** 31.941   **3.** 26.59134

4. 0.477   **5.** 10.2854   **6.** 121.3246

Write the following numbers as words.

7. 45.354   **8.** 16.75

Write the following words as numbers.

9. Nineteen and eight tenths

10. Fifty and ninety-seven thousandths

Add each of the following.

11. 71.613 + 23.7   **12.** 10.42 + 1.608

13. 123.1 + 4.45   **14.** $13 + $4.89 + $0.62

Subtract each of the following.

15. $72 − $14.37   **16.** 26.7 − 22.67

17. 132.3 − 14.9

Work each word problem.

18. The earth is not perfectly round. At the equator it is 7926.4 miles wide. Measured at the poles it is only 7899.9 miles wide. By how many miles does the earth "bulge" at the equator?

19. How much change should you receive from a $20 bill if you bought something costing $11.63?

20. To find the perimeter of a field, a surveyor adds the measurements of all the sides. Find the perimeter of the field shown below.

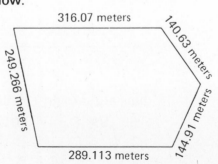

316.07 meters
140.63 meters
249.266 meters
144.91 meters
289.113 meters

Your Answers

1. _____

2. _____

3. _____

4. _____

5. _____

6. _____

7. _____

   _____

   _____

8. _____

   _____

   _____

9. _____

10. _____

11. _____

12. _____

13. _____

14. _____

15. _____

16. _____

17. _____

18. _____

19. _____

20. _____

8 Multiplying and Dividing Decimals

Aims you toward:

- Multiplying decimals and getting the right answer.
- Dividing decimals and getting the right answer.
- Rounding off numbers.
- Multiplying and dividing by multiples of 10.

In this unit you will work exercises like these:

- Multiply 6.78×34.56.
- Divide 4.53 by 90.1.
- Round off 4.532 to the nearest hundredth.
- Multiply 20.83×100.

Multiplying Decimals

All you need to know in order to multiply decimals is how to count DECIMAL PLACES. The number of decimal places is the number of *digits* to the right of the decimal place.

Example How many decimal places are there in 4.683?

There are 3 digits to the *right* of the decimal point, so 4.683 has 3 decimal places.

Here's how to use decimal places when multiplying decimals.

Example Multiply 1.98 × 3.3.

```
  1.9 8      Set up the problem, ignoring the decimal points.
× 3.3        Multiply as you would any multiplication problem.
  5 9 4
5 9 4
6 5 3 4      To figure out where to put the decimal point in your answer,
             count the total number of decimal places in the two
  1.9 8      numbers being multiplied.
× 3.3
  5 9 4      There are 2 decimal places in 1.98, and 1 decimal place in 3.3.
             There must be 3 decimal places in the answer.
5 9 4        Count the decimal places starting from the right.
6.5 3 4      The answer is 6.534.
  3 2 1
```

Does 6.534 make sense as the answer in the above example? 1.98 is approximately 2 and 3.3 is approximately 3. You know that 2 × 3 = 6, so 6.534 is a reasonable answer.

To Multiply Decimals	Multiply, ignoring the decimal points. Count the total number of decimal places in the two numbers being multiplied. Put the decimal point in the answer so that the answer has the same number of decimal places as there are in the two numbers being multiplied.

Sometimes you must add zeros to the left of the answer when you multiply decimals.

Example Multiply 4.34 × 0.004.

```
   4.3 4      Multiply, ignoring the decimal points.
× 0.0 0 4     Count up the total number of decimal places in the problem.
   1 7 3 6    There are 5 decimal places, so there must be 5 decimal
              places in the answer.
  .0 1 7 3 6  Begin counting from the right. Add a zero at the left to get 5
   5 4 3 2 1  decimal places, and put a decimal point in front of the zero.
              The answer is 0.01736.
```

Exercises

Multiply each of the following. Make sure you put the decimal point in the correct place.

1. 1.125 $\times$ 0.03	**2.** 3.201 $\times$0.241	**3.** 0.64 $\times$0.16	**4.** 439 $\times$0.04	**5.** 0.0037 $\times$0.046
6. 0.1127 $\times$0.0106	**7.** 0.043 $\times$ 32	**8.** 11.62 $\times$ 0.37	**9.** 0.698 $\times$0.875	**10.** 21.6 $\times$ 3.8
11. 5.8 $\times$6.3	**12.** 1.75 $\times$ 2.6	**13.** 42.6 $\times$ 7	**14.** 1.407 $\times$ 8	**15.** 23.9 $\times$ 4.8
16. 6.7 $\times$0.009	**17.** .316 $\times$0.07	**18.** 7.8 $\times$0.003	**19.** 0.0321 $\times$ 0.008	**20.** 0.592 $\times$0.048

21. 3.7×62.1 **22.** 1.36×0.8 **23.** 0.53×17.2 **24.** 25.7×1.2

25. What is the gross pay for a musician who earns $45.50 per hour and works 18 hours?

26. Ethyl left work early to shop for her mother's birthday gift. When she punched out at the time clock, it recorded 6.25 hours of work. If she gets paid $9.05 per hour as a technician, how much did Ethyl earn that day?

Answers

26. $56.56

21. 229.77 **22.** 1.088 **23.** 9.116 **24.** 30.84 **25.** $819.00

16. 0.0603 **17.** 0.02212 **18.** 0.0234 **19.** 0.0002568 **20.** 0.028416

11. 36.54 **12.** 4.55 **13.** 298.2 **14.** 11.256 **15.** 114.72

6. 0.00119462 **7.** 1.376 **8.** 4.2994 **9.** 0.610750 **10.** 82.08

1. 0.03375 **2.** 0.771441 **3.** 0.1024 **4.** 17.56 **5.** 0.0001702

Comparing Decimals and Fractions

Decimals and fractions are really the same things. Remember that we began by calling decimals DECIMAL FRACTIONS in Unit 7.

Example $0.3 \times 0.25 = 0.075$

0.3 can be written as $\frac{3}{10}$, and 0.25 as $\frac{25}{100}$.

$$\frac{3}{10} \times \frac{25}{100} = \frac{75}{1000}$$

0.075 is the same as $\frac{75}{1000}$.

Example $2.5 \times 1.25 = 3.125$

Change the decimals to fractions.

$$2.5 = 2\frac{5}{10}, \quad \text{and} \quad 1.25 = 1\frac{25}{100},$$

so we have $2\frac{5}{10} \times 1\frac{25}{100}$.

$$\frac{25}{10} \times \frac{125}{100} = \frac{3125}{1000} = 3\frac{125}{1000}$$

3.125 is also written as $3\frac{125}{1000}$.

Exercises

Here are a few exercises to give you a little practice. The first one is done for you.

First, multiply the problem as you see it, using decimals. Next, change the decimals to fractions and multiply. Compare your two answers.

1. 0.5×0.25

First, $0.5 \times 0.25 = 0.125$.

Next, $\frac{5}{10} \times \frac{25}{100} = \frac{125}{1000}$.

0.125 is the same as $\frac{125}{1000}$.

2. 0.2×0.45

3. 0.75×0.25

4. 1.5×1.25

Answers

1. $0.125 = \frac{125}{1000}$

2. $0.090 = \frac{90}{1000}$

3. $0.1875 = \frac{1,875}{10,000}$

4. $1.875 = 1\frac{875}{1000}$

Dividing Decimals by Whole Numbers

To understand division of decimals, first look at an ordinary division problem with whole numbers.

```
      6
36)216
   216
     0
```

Here is the same problem with a decimal point added to the dividend.

```
      6.
36)216.
   216
     0
```

The important thing to notice is that the decimal point goes in the quotient directly above the decimal point in the dividend.

Example   Divide 3.72 by 4.

```
4)3.72
```
Set up the problem as you normally would.
Put the decimal point in the quotient directly above the decimal point in the dividend.

```
  0.93
4)3.72
  3 6
   12
   12
    0
```
Divide as you normally would.
The answer is 0.93.

Sometimes you need one or more zeros in the quotient in order to keep the decimal point in the correct place.

Example   Divide 0.145 by 5.

```
5)0.145
```
Set up the problem as you normally would.
The decimal point goes directly above the decimal point in the dividend.

```
  0.029
5)0.145
   10
   45
   45
    0
```
Divide as you normally would.
Put a zero in the tenths place to keep the decimal point in the correct place.
The answer is 0.029.

To Divide a Decimal by a Whole Number	Put the decimal point in the quotient directly above the decimal point in the dividend. Divide as you normally do. Use zeros as needed to keep the decimal point in the correct place.

Exercises

1. $5\overline{)33.5}$ 2. $25\overline{)103.25}$ 3. $7\overline{)18.2}$ 4. $38\overline{)198.36}$

5. $2\overline{)1.3174}$ 6. $3\overline{)0.267}$ 7. $9\overline{)0.0432}$ 8. $46\overline{)3.68}$

9. $6\overline{)3.54}$ 10. $11\overline{)0.253}$ 11. $4\overline{)8.64}$ 12. $27\overline{)99.9}$

13. $8\overline{)2.8376}$ 14. $6\overline{)75.36}$ 15. $211\overline{)4937.4}$ 16. $4\overline{)21.56}$

17. $19\overline{)0.665}$ 18. $36\overline{)198.36}$ 19. $47\overline{)0.376}$ 20. $3\overline{)287.7}$

21. $234.5 \div 5$ 22. $494.64 \div 72$ 23. $0.1581 \div 17$ 24. $29.304 \div 22$

25. A newspaper distributor pays the newspaper agency $101.04 for 842 copies. What is her cost per copy?

26. A church group sent $128.64 to its mission post overseas. This covered the cost of feeding 536 children 1 meal each. How much was the cost of 1 meal?

In previous units you used two different ways to write the answer to a division problem when it does not come out exactly. You have learned to write the answer as a whole number and a remainder, or as a whole number and a fraction.

$$
\begin{array}{r}
6 \\
6\overline{)39} \\
\underline{36} \\
3
\end{array}
\qquad
\begin{array}{l}
39 \div 6 = 6,\ \text{remainder 3, or} \\[6pt]
39 \div 6 = 6\dfrac{3}{6} = 6\dfrac{1}{2}
\end{array}
$$

Now see how to write the answer as a decimal.

$$
\begin{array}{r}
6. \\
6\overline{)39.} \\
\underline{36} \\
3
\end{array}
$$
Set up the problem with a decimal point in the dividend.
Divide.
Instead of writing the 3 as a remainder, or as the numerator in a fraction, continue dividing.

$$
\begin{array}{r}
6.5 \\
6\overline{)39.0} \\
\underline{36} \\
3\,0 \\
\underline{3\,0} \\
0
\end{array}
$$
To continue dividing, add a zero to the dividend.
(This does not change the value of the dividend.)
Bring down the zero and divide.

$39 \div 6 = 6.5$

Look at the different ways of expressing the answer to the same division problem. Each way suits a different situation.

When you divide a decimal by a whole number, you can add zeros to the dividend until the answer comes out exactly.

Example   Divide 8 into 4.2.

$$
\begin{array}{r}
0.5 \\
8\overline{)4.2} \\
\underline{4\,0} \\
2
\end{array}
$$
Divide as you normally would.
Remember the decimal point.

$$
\begin{array}{r}
0.52 \\
8\overline{)4.20} \\
\underline{4\,0} \\
20 \\
\underline{16} \\
4
\end{array}
$$
It doesn't come out exactly, so add a zero to the dividend and continue dividing.

$$
\begin{array}{r}
0.525 \\
8\overline{)4.200} \\
\underline{4\,0} \\
20 \\
\underline{16} \\
40 \\
\underline{40} \\
0
\end{array}
$$
It still doesn't come out exactly, so add another zero.
This time it comes out exactly.
The answer is 0.525.

Exercises

Divide, adding zeros as needed until the answer comes out exactly.

1. $8\overline{)59}$

2. $6\overline{)6.9}$

3. $4\overline{)45.4}$

4. $32\overline{)9.12}$

5. $3\overline{)14.82}$

6. $8\overline{)50}$

7. $24\overline{)5.424}$

8. $6\overline{)5.658}$

9. $2\overline{)37}$

10. $42\overline{)254.94}$

11. $52\overline{)13}$

12. $2\overline{)2.3}$

13. $75\overline{)588}$

14. $5\overline{)0.38}$

15. $25\overline{)17.7}$

16. $15\overline{)138.9}$

17. $50\overline{)129.4}$

18. $2\overline{)79.3}$

19. $5\overline{)0.1357}$

20. $5\overline{)1547}$

21. $44\overline{)1.65}$

22. $25\overline{)4.84}$

23. $3.78 \div 375$

24. $51.8 \div 8$

25. A fashion designer bought 25 yards of fabric for a total of $408. What was the cost of the fabric per yard?

26. Russ lives in a large city but enjoys working on a farm each summer. One day he loaded 625 bales of hay and earned $50. How much did he earn per bale?

Answers

25. $16.32 per yard

26. $0.08, or 8 cents

1. 7.375
2. 1.15
3. 11.35
4. 0.285
5. 4.94
6. 6.25
7. 0.226
8. 0.943
9. 18.5
10. 6.07
11. 0.25
12. 1.15
13. 7.84
14. 0.076
15. 0.708
16. 9.26
17. 2.588
18. 39.65
19. 0.02714
20. 309.4
21. 0.0375
22. 0.1936
23. 0.01008
24. 6.475

Rounding Off Decimals

Numbers are often rounded off to make them easier to work with or easier to understand. A decimal like 17.89236487 is pretty clumsy. You can round off 17.89236487 to the nearest tenth and get 17.9. This same number rounded off to the nearest hundredth is 17.89. To the nearest thousandth it is 17.892. When you round off a number, you must specify to which place it is being rounded off.

To Round Off Decimals	Drop all the digits to the right of the place to which you are rounding off. If the first digit you drop is less than 5, the part that remains is not changed. If the first digit you drop is greater than or equal to 5, raise the preceding digit by 1.

Example   Round off 3.421 to the nearest tenth.

$$3.42\cancel{1}$$
$$3.4$$

Drop the digits to the right of the tenths place. The first digit you drop is 2. Since 2 is less than 5, nothing else changes.

Another Example   Round off 0.7064 to the nearest hundredth.

$$0.70\cancel{64}$$
$$0.71$$

Drop the digits to the right of the hundredths place. The first digit you drop is 6. Since 6 is greater than 5, you increase the preceding digit by 1. Here you increase 0 to 1.

Sometimes the process is described as rounding off to a certain number of *decimal places*. The process is the same; it is just another way of specifying the place to which you are rounding off. For example, "round off to three decimal places" is the same as "round off to the nearest thousandth."

You can also round off whole numbers, or round off decimals to get whole numbers.

Example   Round off 2681.77 to the nearest hundred.

$$2681.\cancel{77}$$
$$2700$$

Drop the decimal point and the digits in the tenths and hundredths places.
Replace the digits in the tens and ones places with zeros. The first digit you replace is greater than 5, so increase the preceding digit by 1. Here you increase 6 to 7.

Whenever you work with money, the answer must be rounded off to the nearest cent. You cannot express an amount of money as $1.783. When you purchase something, the amount is usually rounded to the nearest cent. So $1.783 is written as $1.78.

When working problems with decimals, you are often asked to round off the answers.

Example   Multiply 3.523 by 75.2. Round off your answer to the nearest hundredth.

3.5 2 3	Multiply as you normally would.
× 7 5.2	The answer must have 4 decimal places.
7 0 4 6	
1 7 6 1 5	
2 4 6 6 1	
2 6 4.9 2 9 6	Now round off the answer to the nearest hundredth.

264.929̶6̶   Drop the digits to the right of the hundredths places.

2 6 4.9 3   9 is greater than 5, so raise the 2 to a 3.
The answer is 264.93.

To Round Off Products	Multiply as usual. Round off the product using the rules on page 155.

The division problems you have worked so far have eventually come out exactly if you added enough zeros to the dividend. There are some division problems that come out exactly only after adding many, many zeros. Other division problems will never come out exactly, no matter how many zeros you add. In both of these cases, you need to round off the quotient.

Here's an example of a division problem that does not come out exactly.

Example   Divide 58 into 1.98. Round off to the nearest ten-thousandth.

$$
\begin{array}{r}
.03413 \\
58\overline{)1.98000} \\
\underline{1.74} \\
240 \\
\underline{232} \\
80 \\
\underline{58} \\
220 \\
\underline{174} \\
46
\end{array}
$$

Set up the problem and divide as usual.
Keep adding zeros and dividing until the quotient gets to the hundred-thousandth place.
To round off to the ten-thousandth place, drop the last 3.
3 is less than 5, so nothing else changes.
The answer is 0.0341.

If you had not been told to round off your answer in the last example, it would have gone on for quite a while.

To Round Off Quotients	Divide the problem until the quotient gets to the place one beyond the place to which you are rounding off. Round off the quotient using the rules on the previous page.

Work the exercises on the next page.

Exercises

Round off each decimal to the nearest tenth.

1. 2.3094   **2.** 1.56   **3.** 0.072   **4.** 0.35   **5.** 47.193

Round off each decimal to the nearest hundredth.

6. 5.327   **7.** 0.6194   **8.** 12.053   **9.** 6.11749   **10.** 95.1302

Round off each decimal to the nearest thousandth.

11. 0.00367   **12.** 5.1922   **13.** 0.2847   **14.** 6.1008   **15.** 0.42831

Multiply each of the following. Round off your answers to the nearest hundredth.

16. 0.1051	**17.** 8.01	**18.** 3.03	**19.** 1.50	**20.** 0.124
$\times$ 0.6	$\times$ 6.4	$\times$ 0.28	$\times$ 0.32	$\times$ 1.9

Divide each of the following. Round off your answers to the nearest thousandth.

21. $19\overline{)5.0162}$ **22.** $52\overline{)73.2}$ **23.** $6\overline{)37.5162}$ **24.** $14\overline{)6.501}$

25. $7\overline{)0.44336}$ **26.** $3\overline{)1.79}$ **27.** $73\overline{)4.08}$ **28.** $28\overline{)9.2}$

29. If you purchase 0.8 of a pound of peanuts at $2.39 per pound, how much must you pay?

30. In 1983 the United States population consumed 13,956 million pounds of chicken. There were 232.6 million people living in the United States at that time. How many pounds of chicken were eaten per person?

Answers

30. 60 pounds
29. 2.39 $\times$ 0.8 = $1.91
28. 0.329   **27.** 0.056   **26.** 0.597
25. 0.063   **24.** 0.464   **23.** 6.253   **22.** 1.408   **21.** 0.264
20. 0.24   **19.** 0.48   **18.** 0.85   **17.** 51.26   **16.** 0.06
15. 0.428   **14.** 6.101   **13.** 0.285   **12.** 5.192   **11.** 0.004
10. 95.13   **9.** 6.12   **8.** 12.05   **7.** 0.62   **6.** 5.33
5. 47.2   **4.** 0.4   **3.** 0.1   **2.** 1.6   **1.** 2.3

Repeating Decimals

Sometimes when you divide and express the answer as a decimal, the quotient is an endless string of repeating numbers.

Example   Divide 3.8 by 6.

$$
\begin{array}{r}
0.6333 \\
6\overline{)3.8000} \\
\underline{3\,6} \\
20 \\
\underline{18} \\
20 \\
\underline{18} \\
2
\end{array}
$$

This quotient goes on forever with an endless string of threes.

It is called a REPEATING DECIMAL.

Obviously you can't write out the entire answer to a problem like the one above. There are two things you can do. You can round off the answer using the methods you just learned, or you can use a symbol to show that the 3 repeats indefinitely.

This symbol is a bar that goes above the repeating number or numbers in a repeating decimal. The quotient in the above example would be written $0.6\overline{3}$.

Examples   78.272727 . . . is written $78.\overline{27}$.

191.66666 . . . is written $191.\overline{6}$.

3.037037037 . . . is written $3.\overline{037}$.

0.16666 . . . is written $0.1\overline{6}$.

One way to determine quickly that the quotient of a division problem will be a repeating decimal is to look for remainders that are the same.

Example   Divide 41 by 33.

$$
\begin{array}{r}
1.24 \\
33\overline{)41.00} \\
\underline{33} \\
80 \\
\underline{6\,6} \\
1\,40 \\
\underline{1\,32} \\
8
\end{array}
$$

Divide, using a decimal point.

The remainder is the same (8) in two different places. This means that you will keep getting 24 over and over in the quotient. The answer is $1.\overline{24}$.

If you don't understand how I knew that the answer to the last example would be a repeating decimal, continue dividing it out. Notice that you keep getting remainders of 8 and 14.

Looking for remainders that are the same may save you time when dividing.

Exercises

Express each of the following decimals using the bar notation for repeating decimals.

1. 19.3333 …

2. 179.296666 …

3. 3.2486272727 …

4. 56.89898989 …

5. 230.013013013 …

Divide each of the following. Write the quotient as a repeating decimal using the bar notation.

6. 12⟌433.6

7. 3⟌9.2

8. 1.99 ÷ 33

9. 12⟌100

10. 22⟌1.38

11. 3⟌1.79

12. 66⟌86.4

13. 6⟌0.1

14. 22⟌17.8

15. 21⟌112

16. 3⟌4

17. 777⟌233.8

18. 36⟌1.52

19. 0.082 ÷ 11

20. 37 ÷ 6

Answers

1. 19.$\overline{3}$ 2. 179.29$\overline{6}$ 3. 3.2486$\overline{27}$ 4. 56.$\overline{89}$ 5. 230.$\overline{013}$
6. 36.1$\overline{3}$ 7. 3.0$\overline{6}$ 8. 0.060$\overline{3}$ 9. 8.$\overline{3}$ 10. 0.0627
11. 0.596   12. 1.3$\overline{09}$ 13. 0.01$\overline{6}$ 14. 0.80$\overline{9}$ 15. 5.$\overline{3}$
16. 1.$\overline{3}$ 17. 0.300$\overline{9}$ 18. 0.042   19. 0.00745   20. 6.1$\overline{6}$

Multiplying and Dividing by Multiples of 10

Many times you need to multiply or divide numbers by multiples of 10: 10, 100, 1000, and so on. There is a way to multiply or divide numbers by multiples of 10 simply by moving the decimal point.

To multiply a number by 10, move the decimal point 1 place to the *right.*

$$3.4 \times 10 = 34. \quad \text{or} \quad 34$$
$$198.56 \times 10 = 1,985.6$$

To multiply a number by 100, move the decimal point 2 places to the right.

$$567.9 \times 100 = 56,790$$
$$45.8932 \times 100 = 4,589.32$$

To Multiply a Number by a Multiple of 10	Count the number of zeros there are in the multiple of 10. Move the decimal point that many places to the RIGHT.

Example   Multiply 56.98 by 10,000.

$$56.98 \times 10,000 = 569,800 \qquad \text{10,000 has 4 zeros, so move the}$$
decimal point 4 places to the right.

To divide a number by 10, move the decimal point 1 place to the *left.*

$$54.98 \div 10 = 5.498 \qquad 3.28 \div 10 = 0.328$$

To divide a number by 100, move the decimal point 2 places to the left.

$$6.78 \div 100 = 0.0678 \qquad \text{Notice that we must add a zero to put the}$$
decimal point in the correct place.
$$543.7 \div 100 = 5.437$$

To Divide a Number by a Multiple of 10	Count the number of zeros there are in the multiple of 10. Move the decimal point that many places to the LEFT.

Why does this work? Think back to naming place values for decimals. Each place is one tenth of the one to the left of it. So when you multiply or divide by multiples of 10, you are changing the place values of each of the digits in the decimal. To see that it works, do each of the above problems the long way. Compare your answers to the ones above.

Exercises

Find the answer to each of the following problems without working them out the long way.

1. 100 × 2.048 **2.** 0.78 × 1000 **3.** 4.7 × 100 **4.** 6300 ÷ 100

5. 6 ÷ 1000 **6.** 8.4 ÷ 100 **7.** 9.533 × 10 **8.** 10 × 26.955

9. 0.7 × 1000 **10.** 12.5 ÷ 10 **11.** 100 × 5.6 **12.** 0.32 ÷ 10

13. 9.33 ÷ 10 **14.** 801.6 ÷ 100 **15.** 1.7 ÷ 100 **16.** 10 × 0.791

17. 0.3 × 100 **18.** 152 ÷ 10 **19.** 306 ÷ 100 **20.** 45 ÷ 100

21. There are 1,000 grams in 1 kilogram. How many grams are there in 237 kilograms?

22. The Watertown Park District sold 10,000 lottery tickets for $1.50 each. How much money did the Park District get from the sale of the tickets?

23. Dynacolor Film Processors estimates that they will need to spend a total of $13,172,543 on new machinery. They plan to spread the cost of this machinery out evenly over the next 10 years. How much will they need to spend each year?

24. An electron microscope magnifies objects so that they look 1,000 times as big as their actual size. If a body cell measures 0.004 inches long, how long would it seem to be under the microscope?

Answers

24. 4 inches long
21. 237,000 grams **22.** $15,000 **23.** $1,317,254.30
16. 7.91 **17.** 30 **18.** 15.2 **19.** 3.06 **20.** 0.45
11. 560 **12.** 0.032 **13.** 0.933 **14.** 8.016 **15.** 0.017
6. 0.084 **7.** 95.33 **8.** 269.55 **9.** 700 **10.** 1.25
1. 204.8 **2.** 780 **3.** 470 **4.** 63 **5.** 0.006

Dividing Decimals by Decimals

So far you have been dividing decimals by whole numbers. What happens when you want to divide a decimal by another decimal? Take, for example, 4.32 ÷ 0.9.

$$0.9\overline{)4.32}$$ Set up the problem as usual.

$$0\,9.\overline{)4\,3.2}$$ Before you divide, move the decimal point in the divisor 1 place to the right to make a whole number.
Also move the decimal point in the dividend 1 place to the right.

$$
\begin{array}{r}
4.8 \\
9.\overline{)4\,3.2} \\
\underline{3\,6} \\
7\,2 \\
\underline{7\,2} \\
0
\end{array}
$$

Ignore the old positions of the decimal points.
Now that you have a decimal divided by a whole number, you can divide easily.
The decimal point goes in the answer directly above the new position of the decimal point in the dividend.

To Divide a Decimal by a Decimal	Move the decimal point in the divisor to the right of the last digit. Move the decimal point in the dividend the same number of places. Divide, putting the decimal point in the quotient above the new position of the decimal point in the dividend.

Another Example   Divide 0.25 into 1.5 following the rules above.

$$0\,25.\overline{)1\,50.}$$ Move the decimal point in the divisor 2 places to the right.
Move the decimal point in the dividend 2 places to the right. (Add a zero.)

$$
\begin{array}{r}
6. \\
25\overline{)1\,50.} \\
\underline{1\,50} \\
0
\end{array}
$$

Divide.

Another Example

This kind of division problem is the trickiest type of all!

Divide 12 by 1.644. Round off the answer to the nearest thousandth.

Set up the problem.

$$1.644 \overline{\smash{)}12}$$

Did you notice that the 12 has no decimal point?

Yes it has!

Think of money: $12 = $12

If you don't see a decimal point, always put it behind the number before you start to work out the problem.

$$1\,644. \overline{\smash{)}12\,000.}$$

Move the decimal point in the divisor 3 places to the right.
Move the decimal point in the dividend 3 places to the right. (Add 3 zeros.)

```
          7.2992
1644)12000.0000
     11508
       492 0
       328 8
       163 20
       147 96
        15 240
        14 796
         4440
         3288
         1152
```

Divide. Add as many zeros as you need.
(We need 4 decimal places so we can round the answer to the nearest thousandth.)

You can't go on dividing forever, so just leave the remainder dangling in the wind.

Round off your answer to the nearest thousandth.

$$7.2992 = 7.299$$

Exercises

Round off the answers to the nearest thousandth if needed.

1. $0.34\overline{)0.238}$ 2. $4.8\overline{)24}$ 3. $6.4\overline{)0.384}$ 4. $1.62\overline{)8}$

5. $12.6\overline{)0.4}$ 6. $0.104\overline{)5}$ 7. $0.8\overline{)0.64}$ 8. $9.2\overline{)4}$

9. $0.13\overline{)0.741}$ 10. $0.2\overline{)0.8}$ 11. $0.72\overline{)6.192}$ 12. $0.03\overline{)0.372}$

13. $3.3\overline{)0.0264}$ 14. $0.006\overline{)0.0588}$ 15. $0.48\overline{)182.4}$ 16. $0.09416 \div 0.4$

17. $0.0884 \div 0.26$ 18. $0.1197 \div 5.7$ 19. $93 \div 6.2$

20. $65.1 \div 9.3$ 21. $0.423 \div 0.009$ 22. $81 \div 1.5$

23. $0.0273 \div 9.1$ 24. $0.1856 \div 0.032$ 25. $0.18036 \div 0.06$

26. To find the assessed value of a piece of real estate, you divide the amount of taxes by the tax rate. For a certain piece of property, taxes are $4,225 and the tax rate is 0.169. What is the assessed value?

27. A jeweler used $91 worth of gold to make a wedding band. If gold was worth $350 per ounce, how many ounces of gold did he use?

28. As the tide rises a total of 26 feet in the Bay of Fundy, sailors must remember to loosen the anchor ropes every hour or the ropes would snap, setting the boats adrift. If the ropes are loosened 6.5 feet every hour, how many hours does it take for the tide to rise?

Answers

26. $25,000 27. 0.26 ounces 28. 4 hours
25. 3.006 24. 5.8 23. 0.003 22. 54 21. 47
20. 7 19. 15 18. 0.021 17. 0.34 16. 0.235
15. 380 14. 9.8 13. 0.008 12. 12.4 11. 9.6
10. 4 9. 5.7 8. 0.435 7. 0.8 6. 48.077
5. 0.032 4. 4.938 3. 0.06 2. 5 1. 0.7

Changing Fractions to Decimals

In the last unit, you learned to change decimals to fractions. Now you will learn to change fractions to decimals.

We have mentioned several times that something like $\frac{4}{5}$ means division.

$$\frac{4}{5} = 5\overline{)4}$$

So to change a fraction to a decimal, divide the numerator by the denominator.

Example   Write $\frac{7}{8}$ as a decimal.

$$
\begin{array}{r}
0.875 \\
8\overline{)7.000} \\
\underline{6\,4} \\
60 \\
\underline{56} \\
40 \\
\underline{40} \\
0
\end{array}
$$

Divide the numerator by the denominator.
Use the decimal point in the dividend and the quotient.

Another Example   Write $\frac{1}{6}$ as a decimal.

$$
\begin{array}{r}
0.16 \\
6\overline{)1.00} \\
\underline{6} \\
40 \\
\underline{36} \\
4
\end{array}
$$

Divide.
You get a remainder of 4 twice, so you know that the answer is a repeating decimal.
Write the answer as $0.1\overline{6}$.

Another Example   Write $17\frac{3}{5}$ as a decimal.

$$
\begin{array}{r}
0.6 \\
5\overline{)3.0} \\
\underline{3\,0} \\
0
\end{array}
$$

Divide the numerator by the denominator.
The whole number does not need to be changed.
The answer is 17.6.

Now try some exercises on changing fractions to decimals.

Exercises

Write each of the following as a decimal. Divide each fraction until the remainder is zero.

1. $\frac{1}{4}$
2. $\frac{3}{4}$
3. $4\frac{7}{10}$
4. $16\frac{1}{20}$
5. $\frac{1}{2}$

6. $31\frac{4}{5}$
7. $\frac{9}{10}$
8. $\frac{17}{100}$
9. $3\frac{1}{80}$
10. $14\frac{7}{16}$

Write each of the following as a decimal. Round off each answer to the nearest thousandth.

11. $\frac{1}{3}$
12. $\frac{1}{9}$
13. $4\frac{2}{3}$
14. $76\frac{1}{15}$
15. $14\frac{5}{6}$

16. $173\frac{3}{7}$
17. $\frac{9}{11}$
18. $\frac{1}{18}$
19. $2\frac{5}{9}$
20. $\frac{6}{7}$

Write each of the following as a decimal. Use the bar notation for repeating decimals. Round off nonrepeating decimals to the nearest thousandth.

21. $\frac{17}{18}$
22. $\frac{5}{21}$
23. $4\frac{3}{17}$
24. $11\frac{13}{19}$
25. $\frac{2}{7}$

26. $\frac{7}{12}$
27. $6\frac{4}{11}$
28. $\frac{2}{9}$
29. $\frac{5}{12}$
30. $\frac{4}{15}$

Answers

1. 0.25	**2.** 0.75	**3.** 4.7	**4.** 16.05	**5.** 0.5
6. 31.8	**7.** 0.9	**8.** 0.17	**9.** 3.0125	**10.** 14.4375
11. 0.333	**12.** 0.111	**13.** 4.667	**14.** 76.067	**15.** 14.833
16. 173.429	**17.** 0.818	**18.** 0.056	**19.** 2.556	**20.** 0.857
21. 0.94	**22.** 0.238	**23.** 4.176	**24.** 11.684	**25.** 0.286
26. 0.583	**27.** 6.36	**28.** 0.2	**29.** 0.416	**30.** 0.26

Multiplying Decimals by Fractions

(This is an optional topic. Ask your teacher about doing this page.)

To multiply decimals by fractions you could change the fractions to decimals or the decimals to fractions, and then work the problems as you have learned. There is a way, however, to work such problems without making these changes.

Recall that you can put any number over 1 in a fraction and not change the value of the number. Use this fact to multiply decimals by fractions.

Example   Multiply $\frac{2}{3} \times 3.78$.

$$\frac{2}{3} \times \frac{3.78}{1}$$ Write 3.78 as a fraction with 1 as the denominator.

$$\frac{2}{\underset{1}{3}} \times \frac{\overset{1.26}{\cancel{3.78}}}{1} = 2.52$$ Now multiply as you would any two fractions. The answer is 2.52.

Another Example   Multiply $\frac{1}{8} \times 4.78$. Round off the answer to 3 decimal places.

$$\frac{1}{\underset{4}{8}} \times \frac{\overset{2.39}{\cancel{4.78}}}{1} = \frac{2.39}{4}$$ Write 4.78 as a fraction.
Multiply.

$$\begin{array}{r} 0.5975 \\ 4\overline{)2.3900} \\ \underline{2.0} \\ 39 \\ \underline{36} \\ 30 \\ \underline{28} \\ 20 \\ \underline{20} \\ 0 \end{array}$$

Divide to change the improper fraction to a decimal.
Round off 0.5975 to 3 decimal places.
The answer is 0.598.

Another Example   Multiply $3\frac{5}{6} \times 2.745$. Round off the answer to 3 decimal places.

$$3\frac{5}{6} = \frac{23}{6}$$ Change $3\frac{5}{6}$ to an improper fraction and multiply.

$$\frac{23}{\underset{2}{6}} \times \frac{\overset{0.915}{\cancel{2.745}}}{1} = \frac{21.045}{2} = 10.5225 \quad \text{or} \quad 10.523$$

Exercises

Multiply each of the following. Round off to the nearest hundredth if needed.

1. $\frac{3}{4} \times 3.64$ 2. $\frac{2}{5} \times 22.25$ 3. $8.26 \times \frac{1}{3}$ 4. $12.28 \times 1\frac{1}{4}$

5. $9.04 \times 3\frac{1}{2}$ 6. $4.2 \times 9\frac{5}{6}$ 7. $\frac{2}{3} \times 3.192$

8. $7\frac{1}{2} \times 11.2$ 9. $0.888 \times 6\frac{3}{4}$ 10. $4.96 \times \frac{7}{8}$

11. $25.68 \times 3\frac{3}{8}$ 12. $7.56 \times 2\frac{2}{3}$ 13. $46.1 \times \frac{1}{5}$

14. If finished mahogany costs \$4.36 per board foot, how much does $6\frac{1}{2}$ board feet cost?

15. A hospital aide earns \$4.54 per hour. How much does she make in a $7\frac{1}{2}$-hour day?

16. How far could an army missile traveling 1,830 miles per hour go in $4\frac{1}{2}$ minutes?

Answers

Step 2: Multiply $4\frac{1}{2}$ by 30.5: $\frac{\overset{1}{\cancel{9}}}{2} \times \frac{\overset{15.25}{30.5}}{1} = 137.25$ miles.

$1830 \div 60 = 30.5$ miles per minute.

16. Step 1: Convert miles per hour to miles per minute:

11. 86.67	12. 20.16	13. 9.22	14. \$28.34	15. \$34.05
6. 41.30	7. 2.13	8. 84.00	9. 5.99	10. 4.34
1. 2.73	2. 8.90	3. 2.75	4. 15.35	5. 31.64

Word Problems

1. The moon weighs about $\dfrac{1}{81.31}$ as much as the earth. The earth is thought to weigh about 5935.63 trillion tons. How many trillion tons does the moon weigh?

2. Light takes about 1.28 seconds to travel from the earth to the moon. This distance is about 238,856 miles. How fast is the speed of light? (Round off your answer to the nearest whole number.)

3. In 1854, the sailing ship "Flying Cloud" went from New York around South America to San Francisco, a distance of 15,091 miles. Her average speed (measured by dragging a log behind the ship) was 7.07 miles per hour. About how many days did this voyage take? (24 hours = 1 day)

4. Compare your answer in problem 3 to modern travel: in 1983, a passenger train went from Wilmington to Baltimore, a distance of 68.4 miles, in 41 minutes. About how fast was the train traveling in miles per hour? (60 minutes = 1 hour)

Answers

1. $\dfrac{1}{81.31} \times \dfrac{5935.63}{1} = 73$ trillion tons   2. 186606 or 187000 miles per second   3. About 89 days   4. About 100 miles per hour

Unit 8 Test

Use a separate piece of paper to work out each problem.

Multiply or divide as indicated. Round off the answers to the nearest thousandth as needed.

1. 39.6 × 4.2

2. 24.84 ÷ 6

3. 20.865 ÷ 5

4. 42.1 × 3.9

5. 4)‾14.7389‾

6. 375.429 ÷ 12

7. 47.6 × 9.3

8. 43.8 × 2.04

9. 9 ÷ 17

10. 4.987 ÷ 2.6

11. 0.042 × 3.2

12. 5.4213 ÷ 3.42

13. 21.7 × 0.431

14. 0.52)‾7.6351‾

15. 9.06)‾428.17‾

16. 0.0408 × 0.06

17. 1976.2 × 0.005

18. 0.0139 × 4280

19. What is the total cost of 21 truck tires at $86.42 per tire?

20. Suppose you drove 346.2 miles on 16.3 gallons of gas. How many miles per gallon did you get? Round off to the nearest hundredth.

21. The manufacturers of phonograph equipment must make sure that their turntables rotate at precisely the correct speed. If a turntable is to rotate at the rate of 33.33 revolutions per minute, how many revolutions should it make in one second? Round off to the nearest thousandth.

22. A weather watcher timed a lightning flash with a stopwatch. If the speed of sound was 1100 feet per second that day, and if the sound of the thunder took 3.25 seconds to reach him, how far away was the storm?

Your Answers

1. _____

2. _____

3. _____

4. _____

5. _____

6. _____

7. _____

8. _____

9. _____

10. _____

11. _____

12. _____

13. _____

14. _____

15. _____

16. _____

17. _____

18. _____

19. _____

20. _____

21. _____

22. _____

Review Test

Use a separate piece of paper to work out each problem.

Give the place value for the underlined number in each of the following problems.

1. 4618.329
2. 7192.847
3. 2.4936

4. 7005.297
5. 1306.3948
6. 8165.702

Work each of the following problems. Watch out for the signs.

7. 32.4 + 18 + 0.6
8. 6.1 − 2.48

9. 30 − 14.099
10. 8 + 2.8 + 10.02

11. At the end of each day, Ken puts his loose change in a jar. If he has 5 nickels, 9 quarters, 12 pennies, and 6 dimes, how much money does he have altogether?

12. Geraldine had 63 dollars in her checking account, and then she wrote a check for $18.85 for a new purse. How much was left in her account?

Work each of the following problems.

13. 28.4 × 0.74
14. 10.344 ÷ 12

15. 244 ÷ 0.16
16. 65.8 × 76.9

17. 3.014 × 0.036
18. 152.36 ÷ 58.6

19. 5 ÷ 16
20. 0.4968 × 0.000

21. A machinist cut a brass rod 25.65 inches long into pieces that measured 4.275 inches each. How many pieces did he get?

22. Silver is worth about $7.95 per ounce. How much would the silver be worth in a bracelet containing 0.62 ounces of silver?

Your Answers

1. _____
2. _____
3. _____
4. _____
5. _____
6. _____
7. _____
8. _____
9. _____
10. _____
11. _____
12. _____
13. _____
14. _____
15. _____
16. _____
17. _____
18. _____
19. _____
20. _____
21. _____
22. _____

9 | Percent

Aims you toward:

- Knowing the meaning of percent.
- Knowing how to change fractions and decimals to percents.
- Knowing how to change percents to fractions and decimals.

In this unit you will work exercises like these:

- Write $\frac{3}{5}$ as a percent.
- Write 0.98 as a percent.
- Write 34% as a fraction.
- Write 67% as a decimal.

The Meaning of Percent

If Canada's population increased from 23,162,220 to 24,976,820, it would have increased by 1,814,600. But these numbers don't mean very much. They are too large to imagine.

We can say instead that Canada's population increased by 8 PERCENT. An 8 percent population increase means that there are 8 new residents for every 100 old residents.

8 per 100 is 8 percent.

The symbol for percent is %.

The population increase in Canada is 8%.

Percent is the number of parts per hundred. ("Percent" comes from the Latin word "centum," which means 100.)

Example Think of an office with 100 employees. If there are 63 women, then 63% of the employees are women. In this office there would be 37 men. Or, 37% of the employees are men.

Another Example If an apartment building has 100 apartments and 89 of them are rented, what percent of the apartments are rented?

Percent means parts per hundred. There are 89 rented apartments per 100 apartments. So 89% of them are rented.

To check your understanding of percent, work the exercises on the next page.

Exercises

A farmer expects to harvest 100 bushels of corn per acre. What percent of the expected amount would he get if he harvested the following amounts per acre?

1. 87

2. 93

3. 46

4. 88

5. 105

6. 100

7. 95

8. 110

Assume that there are 100 freight cars on a train. Express each of the following as a percent of the total number of cars.

9. 22 of the cars contain grain.

10. 13 of the cars contain large machinery.

11. 11 of the cars carry automobiles.

12. 6 of the cars are empty.

13. A 100-year-old woman said that she spent 58% of her life working. How many years did she work?

14. If 46% of working people travel more than 20 minutes to work, how many workers out of a hundred would you expect to travel more than 20 minutes to work?

15. The weather forecaster says that there is a 60% chance of rain tomorrow. Out of a hundred possible weather conditions, how many would include rain?

16. A manufacturer of kitchen appliances brags its products have only a 2% defect rate. How many defective appliances might there be in 100 appliances?

Answers

1. 87%	**2.** 93%	**3.** 46%	**4.** 88%
5. 105%	**6.** 100%	**7.** 95%	**8.** 110%
9. 22%	**10.** 13%	**11.** 11%	**12.** 6%
13. 58 years	**14.** 46 workers	**15.** 60	**16.** 2 appliances

Changing Numbers to Percents

To change any number to a percent, multiply by 100 and add the % sign.

Example   Change $\frac{3}{4}$ to a percent.

$$\frac{3}{\underset{1}{\cancel{4}}} \times \frac{\overset{25}{\cancel{100}}}{1} = \frac{3 \times 25}{1 \times 1} = \frac{75}{1} = 75 \qquad \text{Multiply by 100.}$$

$$\frac{3}{4} = 75\% \qquad \text{Add the \% sign.}$$

Here's an explanation.

Percent means parts per hundred, so write $\frac{3}{4}$ as a fraction with 100 as the denominator.

$$\frac{3}{4} = \frac{75}{100}$$

In $\frac{75}{100}$ there are 75 parts per 100, so it is equal to 75%.

Another Example   Write 3 as a percent.

$$3 \times 100 = 300$$
$$3 = 300\%$$

Another Example   Express 0.625 as a percent.

$$0.625 \times 100 = 62.5$$
$$0.625 = 62.5\%$$

Another Example   Change $12\frac{1}{3}$ to a percent. Express the answer as a mixed number.

$$12\frac{1}{3} = \frac{37}{3}$$

$$\frac{37}{3} \times \frac{100}{1} = \frac{3700}{3} = 1233\frac{1}{3} \qquad \begin{array}{l}\text{3700} \div \text{3 doesn't divide exactly, so} \\ \text{write the answer as a mixed} \\ \text{number.}\end{array}$$

$$12\frac{1}{3} = 1233\frac{1}{3}\%$$

When a percent, such as the one in the last example, does not come out exactly, either round it off or write it as a mixed number.

Exercises

Change each of the following numbers to percents.

1. $\frac{3}{5}$

2. 0.6

3. $\frac{3}{4}$

4. 0.75

5. $\frac{1}{5}$

6. 0.2

7. 4.06

8. $2\frac{1}{2}$

9. 1

10. $\frac{1}{4}$

11. 0.025

12. $\frac{7}{8}$

13. 0.875

14. $\frac{4}{5}$

15. 0.8

16. $\frac{1}{2}$

17. 0.5

18. $\frac{1}{8}$

19. 0.166

20. 0.50

21. $\frac{1}{10}$

22. 0.64

23. $\frac{1}{6}$

24. 0.04

25. $\frac{3}{8}$

26. Tricky question number one: If $\frac{2}{3}$ of a company's bookkeeping is computerized, what percent of the bookkeeping is *not* computerized?

Answers

1. 60%
2. 60%
3. 75%
4. 75%
5. 20%
6. 20%
7. 406%
8. 250%
9. 100%
10. 25%
11. 2.5%
12. 87.5% or $87\frac{1}{2}$%
13. 87.5%
14. 80%
15. 80%
16. 50%
17. 50%
18. 12.5% or $12\frac{1}{2}$%
19. 16.6%
20. 50%
21. 10%
22. 64%
23. $16.\overline{6}$% or 16.667%
24. 4%
25. 37.5% or $37\frac{1}{2}$%
26. $\frac{1}{3} = 33.\overline{3}$% or 33.333%

Changing Percents to Fractions or Decimals

To change a percent to a fraction or a decimal, divide by 100 and drop the % sign.

Example   Change 60% to a fraction.

$$60\% = 60 \div 100 = \frac{60}{100} = \frac{3}{5}$$

Another Example   Change 60% to a decimal.

$$60\% = 60 \div 100 = 0.60$$ Remember dividing by 100? Move the decimal point 2 places to the *left.*

Wondering how this works?

Percent means parts per hundred so 60% means 60 parts per hundred or $\frac{60}{100}$.

You know that $\frac{60}{100}$ means dividing 60 by 100. If you want a fraction, simply reduce. If you want a decimal, divide.

Another Example   Change 224% to a fraction.

$$224\% = 224 \div 100 = \frac{224}{100} = 2\frac{6}{25}$$

Another Example   Change 224% to a decimal.

$$224\% = 224 \div 100 = 2.24$$

Another Example   Change $12\frac{3}{4}\%$ to a decimal.

$12\frac{3}{4}\% = 12.75\%$ Keep the 12. Divide 3 by 4 to get 0.75.

$12.75 \div 100 = 0.2175$ To divide by 100, move the decimal point 2 places to the left.

Another Example

Change $12\frac{3}{4}\%$ to a common fraction.

$$12\frac{3}{4} \div \frac{100}{1}$$

$$\frac{51}{4} \times \frac{1}{100}$$ Remember to divide by inverting the second fraction and multiplying.

$$\frac{51}{400}$$ The answer is $\frac{51}{400}$.

Exercises

Change each of the following percents to decimals.

1. 102%	**2.** 26%	**3.** 162%	**4.** 16%	**5.** 212%
6. 136%	**7.** 18%	**8.** 1%	**9.** 6%	**10.** 8%

Change each of the following percents to fractions.

11. 75%	**12.** 48%	**13.** 122%	**14.** 30%	**15.** 98%
16. 5%	**17.** 18%	**18.** 50%	**19.** 40%	**20.** 10%

21. A TV repair service advertises that only 14% of the TV sets it repairs need reservicing. What fraction is this?

22. Suppose that 2% of all new automobiles have to be recalled for defects. What fraction of new cars are defective?

Answers

21. $\frac{7}{50}$ **22.** $\frac{1}{50}$

16. $\frac{1}{20}$ **17.** $\frac{9}{50}$ **18.** $\frac{1}{2}$ **19.** $\frac{2}{5}$ **20.** $\frac{1}{10}$

11. $\frac{3}{4}$ **12.** $\frac{12}{25}$ **13.** $1\frac{11}{50}$ **14.** $\frac{3}{10}$ **15.** $\frac{49}{50}$

1. 1.02   **2.** 0.26   **3.** 1.62   **4.** 0.16   **5.** 2.12

6. 1.36   **7.** 0.18   **8.** 0.01   **9.** 0.06   **10.** 0.08

Changes Between Fractions, Decimals, and Percents

You now know how to make all of the possible changes between fractions, decimals, and percents. Study the six types shown as a review below. Look back at Units 7 and 8 if you need more examples of changes between fractions and decimals.

TO CHANGE FROM A . . .

1. FRACTION TO A DECIMAL: Divide the numerator by the denominator.

$$\frac{1}{4} = 0.25 \qquad 4\overline{)1.00} \quad \begin{array}{r} .25 \\ \hline \end{array}$$

$$\begin{array}{r} \underline{8} \\ 20 \\ \underline{20} \\ 0 \end{array}$$

2. DECIMAL TO A FRACTION: The numerator is the number to the right of the decimal point. The denominator is found by the place value of the right-hand digit.

$$0.42 = \frac{42}{100} = \frac{21}{50}$$

3. FRACTION TO A PERCENT: Multiply by 100. Add the % sign.

$$\frac{3}{\cancel{5}} \times \frac{\cancel{100}^{20}}{1} = \frac{60}{1} = 60$$

$$\frac{3}{5} = 60\%$$

4. DECIMAL TO A PERCENT: Multiply by 100. Add the % sign.

$$0.821 \times 100 = 82.1$$

$$0.821 = 82.1\%$$

5. PERCENT TO A DECIMAL: Divide by 100 and drop the % sign.

$$106\% = 106 \div 100 = 1.06$$

6. PERCENT TO A FRACTION: Divide by 100 and drop the % sign.

$$106\% = 106 \div 100 = \frac{106}{100} = 1\frac{3}{50}$$

You will need to keep all of these procedures in mind when you take the next test. First work the review exercises on the next page.

Review Exercises

Fill in the following blank spaces. Use page 180 as your guide.

	PERCENT	FRACTION	DECIMAL
1.	60%		
2.			0.24
3.		$\frac{4}{5}$	
4.			0.65
5.		$\frac{9}{10}$	
6.	8%		
7.	48%		
8.			0.18
9.		$\frac{6}{12}$	
10.	85%		
11.			0.04
12.		$\frac{1}{5}$	
13.		$\frac{4}{16}$	
14.	100%		

Review Exercises

Fill in the blanks with either a fraction, a decimal, or a percent equal to the given number. Round off your answers to three decimal places if needed.

	FRACTION	DECIMAL	PERCENT
1.			52%
2.	$\frac{3}{7}$		
3.			82%
4.		0.76	
5.	$\frac{2}{9}$		
6.			35%
7.		0.813	
8.		1.42	
9.			105%
10.	$2\frac{2}{3}$		
11.			16%
12.			22%
13.		0.644	
14.	$\frac{1}{2}$		

Answers

1. $\frac{13}{25}$, 0.52   2. 0.429, 42.9%   3. $\frac{41}{50}$, 0.82   4. $\frac{19}{25}$, 76%

5. 0.2, 22.$\overline{2}$%   6. $\frac{7}{20}$, 0.35   7. $\frac{813}{1000}$, 81.3%   8. 1$\frac{21}{50}$, 142%

9. 1$\frac{1}{20}$, 1.05   10. 2.6, 266.$\overline{6}$%   11. $\frac{4}{25}$, 0.16   12. $\frac{11}{50}$, 0.22

13. $\frac{161}{250}$, 64.4%   14. 0.5, 50%

More Review Exercises

Complete the following. Check with page 182 if you must, but try to do as much work as possible without looking back in the book. Don't forget to reduce your fractions to lowest terms.

1. Write $\frac{7}{8}$ as a decimal and as a percent.

2. Change 0.525 to a percent and a fraction.

3. 12.5% is what decimal? What fraction?

4. Write 0.375 as a fraction and a percent.

5. Change $\frac{3}{10}$ to a percent and a decimal.

6. Express 95% as a fraction and a decimal.

7. 34% = _____ and _____
   (decimal)   (fraction)

8. 0.12 = _____% and _____
                            (fraction)

9. $\frac{8}{10}$ = _____ and _____%
         (decimal)

10. 110% = _____ and _____
            (fraction)   (decimal)

Answers

9. 0.8, 80%   10. $1\frac{1}{10}$, 1.10

5. 30%, 0.3   6. $\frac{19}{20}$, 0.95   7. 0.34, $\frac{17}{50}$ 8. 12%, $\frac{3}{25}$

1. 0.875, 87.5%   2. 52.5%, $\frac{21}{40}$ 3. 0.125, $\frac{1}{8}$ 4. $\frac{3}{8}$, 37.5%

Unit 9 Test

Use a separate piece of paper to work out each problem.

Each of the following numbers is written either as a decimal, a fraction, or a percent. Write each number in its other two forms.

1. 316%

2. 3

3. $\dfrac{9}{4}$

4. $2\dfrac{1}{8}$

5. 5.004

6. 3%

7. 100%

8. $1\dfrac{3}{11}$

9. 16%

10. 4

11. 238%

12. $\dfrac{7}{9}$

13. 0.735

14. $2\dfrac{5}{6}$

15. 4.025

16. 1%

17. The Traffic Control Department at Southwest College issued 100 tickets for illegal parking. If 78 of these went to students, what percent of the tickets were for students?

18. The emergency room staff in most community hospitals say that $\dfrac{5}{8}$ of the cases they handle are not true emergencies. What percent of the cases *are* emergencies?

19. To find the sales tax for a purchase, a clerk multiplies the amount of the purchase by 0.06. What percent sales tax is she figuring?

20. Due to improved lighting conditions, the workers on an assembly line increased their output by 12%. What fraction of their normal output was this increase?

Your Answers

1. _____
2. _____
3. _____
4. _____
5. _____
6. _____
7. _____
8. _____
9. _____
10. _____
11. _____
12. _____
13. _____
14. _____
15. _____
16. _____
17. _____
18. _____
19. _____
20. _____

10 | Ratios and Proportions

Aims you toward:

- Knowing the meaning of ratio.
- Knowing the meaning of proportion.
- Knowing how to solve for an unknown term in a proportion.

In this unit you will work exercises like these:

- Do the ratios $\dfrac{5}{6}$ and $\dfrac{7.5}{9}$ make a proportion?

- Find the value of M in this proportion: $\dfrac{3}{4} = \dfrac{M}{6}$.

Ratios

A RATIO is another name for a fraction.

Instead of calling $\frac{2}{3}$ "two thirds," you can call it "the ratio of 2 to 3."

Ratios are fractions that compare two different numbers or amounts. The first number in a ratio should be placed ABOVE the fraction bar and the second number BELOW the bar.

Example   In preparing 10 liters of liquid fertilizer, you need 8 liters of water, and 2 liters of chemical concentrate. What is the ratio of water to concentrate?

The ratio of water to concentrate is

water $\longrightarrow \frac{8}{2}$ (8 to 2).
concentrate $\longrightarrow$

The number for water is placed above the number for concentrate because water was mentioned first in the question.

Since a ratio is a fraction, a ratio can be reduced. The answer to the above example reduced is $\frac{4}{1}$ (4 to 1). You reduce the ratio $\frac{8}{2}$ to $\frac{4}{1}$, NOT to 4. A ratio always compares one number to another number.

Another   The human body burns 250 calories during a 30-minute run. Find the
Example   ratio of calories to minutes. Reduce the ratio to lowest terms.

calories $\longrightarrow \dfrac{250}{30} = \dfrac{25}{3}$
minutes $\longrightarrow$

The ratio in lowest terms is 25 calories to 3 minutes.

Some exercises on ratios are next.

Exercises

Write each of the following ratios in fraction form and reduce to lowest terms.

1. 116 to 50 **2.** 4 to 2 **3.** 18 to 6 **4.** 7 to 49

5. 18 to 1 **6.** 44 to 11 **7.** 11 to 44 **8.** 100 to 2

9. 98 cents to 3 cans **10.** $7.53 to 3 people

11. 60 people to 5 boats **12.** 3 nurses for 750 patients

Find the ratio for each of the following. Reduce to lowest terms.

13. In checking the safety of a building, a fire inspector looks for the ratio of people to fire escapes. If a building has 356 people working in it and 4 fire escapes, what is the ratio?

14. A secretarial service requires that its employees be able to type 57 words per minute. What is the ratio of words to minutes?

15. A small college says that it has a low ratio of professors to students. If the college has 1260 students and 105 professors, what is the ratio of professors to students?

16. At the personnel office of a large company 128 people showed up to apply for 16 job openings. Express this as a ratio of applicants to job openings.

Answers

1. $\frac{58}{25}$ **2.** $\frac{2}{1}$ **3.** $\frac{3}{1}$ **4.** $\frac{1}{7}$

5. $\frac{18}{1}$ **6.** $\frac{4}{1}$ **7.** $\frac{1}{4}$ **8.** $\frac{50}{1}$

9. $\frac{98}{3}$ **10.** $\frac{2.51}{1}$ **11.** $\frac{12}{1}$ **12.** $\frac{250}{1}$

13. $\frac{89}{1}$ **14.** $\frac{57}{1}$ **15.** $\frac{1}{12}$ **16.** $\frac{8}{1}$

Proportions

Two ratios equal to each other form a PROPORTION.

$$\frac{4}{5} = \frac{16}{20}$$

This proportion is read "the ratio of 4 to 5 is equal to the ratio of 16 to 20." It can also be read "4 is to 5 as 16 is to 20."

Everything you learned about equivalent fractions (see Units 5 and 6) is also true for proportions.

To determine that two ratios are equal to each other, you CROSS MULTIPLY.

Example   Do $\frac{6}{9}$ and $\frac{2}{3}$ make a proportion?

$\frac{6}{9} \times \frac{2}{3}$ To cross multiply, multiply 6 × 3 and 9 × 2.

$6 \times 3 = 18$ The cross products equal each other, so the ratios are equal.

$9 \times 2 = 18$ $\frac{6}{9} = \frac{2}{3}$ is a proportion.

Another Example   Do $\frac{7}{5}$ and $\frac{9}{3}$ make a proportion?

$\frac{7}{5} \times \frac{9}{3}$ Cross multiply.

$7 \times 3 = 21$ $\frac{7}{5}$ and $\frac{9}{3}$ do not make a proportion.
$5 \times 9 = 45$

More exercises are on the next page.

Exercises

Decide whether or not each of the following pairs of ratios makes a proportion.

1. $\dfrac{4}{10}$ $\dfrac{14}{35}$

2. $\dfrac{30}{9}$ $\dfrac{40}{12}$

3. $\dfrac{0.6}{0.8}$ $\dfrac{1.5}{2.0}$

4. $\dfrac{15}{10}$ $\dfrac{20}{16}$

5. $\dfrac{4}{7}$ $\dfrac{32}{56}$

6. $\dfrac{3}{8}$ $\dfrac{10}{27}$

7. $\dfrac{40}{160}$ $\dfrac{8}{30}$

8. $\dfrac{8}{30}$ $\dfrac{12}{45}$

9. $\dfrac{1.1}{0.3}$ $\dfrac{44}{12}$

10. $\dfrac{9}{15}$ $\dfrac{81}{130}$

11. $\dfrac{2.8}{4.5}$ $\dfrac{2.1}{3.5}$

12. $\dfrac{0.3}{0.5}$ $\dfrac{3.0}{4.5}$

13. $\dfrac{12}{21}$ $\dfrac{10}{15}$

14. $\dfrac{18}{14}$ $\dfrac{54}{42}$

15. $\dfrac{6}{12}$ $\dfrac{2.5}{5}$

16. A group of people living in the Yukon Territory held a weight pulling contest using dog sledges. One team of 3 dogs pulled 600 pounds. The second team of 4 dogs pulled 800 pounds. The judges rules the contest a tie, saying that both teams pulled equally in proportion to their sizes. Were the judges right?

17. A particular individual weighs 130 pounds and is 65 inches tall. Another individual weighs 120 pounds and is 64 inches tall. If you compare weights to heights, do you get a proportion?

18. A 5-ton truck used 100 gallons of fuel to make a trip. Another truck weighing 2 tons used 40 gallons of fuel to make the same trip. The owner said that for their sizes both trucks did equally well. Was he right? (Do the two ratios make a proportion?)

Answers

1. Yes	2. Yes	3. Yes	4. No	5. Yes	6. No
7. No	8. Yes	9. Yes	10. No	11. No	12. No
13. No	14. Yes	15. Yes	16. Yes	17. No	18. Yes

Solving a Proportion for an Unknown Term

The four numbers in a proportion are called TERMS. For example, in the proportion

$\dfrac{6}{20} = \dfrac{9}{30}$, 6, 20, 9, and 30 are all TERMS of the proportion.

Every proportion has four terms. When only three of the terms are known, you can always find the fourth term.

Example   Look at this proportion: $\dfrac{7}{2} = \dfrac{M}{3}$.

One of the terms is unknown. It is represented by the letter "*M.*" You use the letter until you know what number is to go there.

To find the value of *M,* cross multiply. The cross products will equal each other.

$$\dfrac{7}{2} \diagup\!\!\!\!\diagdown \dfrac{M}{3}$$

$$7 \times 3 = 2 \times M$$

Next multiply 7×3.

$$7 \times 3 = 2 \times M$$
$$\underbrace{}$$
$$21 \;\; = 2 \times M$$

Now divide both 21 and $2 \times M$ by 2.

$21 = 2 \times M$ is an EQUATION. An equation is an expression with an $=$ sign. As you know, the part to the left of the $=$ sign must equal the part to the right of the $=$ sign. To keep them equal, whatever operation you perform on one side you must also perform on the other side.

In the equation $21 = 2 \times M$, dividing both sides by 2 gives you

$$\dfrac{21}{2} = \dfrac{2 \times M}{2}.$$

$21 \div 2 = 10.5$, and the 2's in $\dfrac{2 \times M}{2}$ cancel each other, so you have

$$10.5 = M, \text{ or } M = 10.5.$$

This method can be used to find an unknown term in any proportion. The example is shown again on the next page.

Here's the whole thing put together.

$$\frac{7}{2} = \frac{M}{3}$$

$$\frac{7}{2} \begin{matrix} \\ \times \\ \end{matrix} \frac{M}{3}$$

$$7 \times 3 = 2 \times M$$

$$21 = 2 \times M$$

$$\frac{21}{2} = \frac{\cancel{2} \times M}{\cancel{2}}$$

$$10.5 = M$$

To check the answer to this proportion problem, substitute 10.5 for M in the original proportion. Cross multiply to see if the ratios are equal.

$$\frac{7}{2} = \frac{10.5}{3}$$

$$7 \times 3 = 2 \times 10.5$$

$$21 = 21$$

The answer is correct.

Here are a few more examples of how to find the missing term in a proportion.

Another Example Find B in this proportion: $\dfrac{3}{B} = \dfrac{5}{9}$.

$\dfrac{3}{B} = \dfrac{5}{9}$	Cross multiply.
$3 \times 9 = B \times 5$	Multiply 3×9.
$27 = B \times 5$	Divide both sides by 5.
$\dfrac{27}{5} = \dfrac{B \times \cancel{5}}{\cancel{5}}$	$27 \div 5 = 5.4$
	The 5's on the right side cancel.
$5.4 = B$	

Now check the answer by substituting 5.4 for B.

$\dfrac{3}{5.4} = \dfrac{5}{9}$	Cross multiply.
$3 \times 9 = 5.4 \times 5$	
$27 = 27$	The answer is correct, so $B = 5.4$.

Another Example

In the proportion $\dfrac{4}{7} = \dfrac{64}{k}$, find the value of k.

$\dfrac{4}{7} = \dfrac{64}{k}$ Cross multiply to solve for k.

$4 \times k = 7 \times 64$

$4 \times k = 448$

$\dfrac{4 \times k}{4} = \dfrac{448}{4}$

$k = 112$

Another Example

Architects often make models of their proposed building plans. An architect's model is 2 feet high. If the model compares with the actual building by a ratio of 1 to 25, how tall is the actual building?

Set up a proportion. First determine the two ratios.

You know that 1 foot on the model corresponds to 25 feet on the building. So the first ratio is $\dfrac{1}{25}$.

You have a model 2 feet high and want to know how high the building is. Let the height of the building equal B. The second ratio is $\dfrac{2}{B}$. The proportion is $\dfrac{1}{25} = \dfrac{2}{B}$.

Solve this proportion for B.

$\dfrac{1}{25} = \dfrac{2}{B}$

$1 \times B = 25 \times 2$

$1 \times B = 50$

$\dfrac{1 \times B}{1} = \dfrac{50}{1}$

$B = 50$

The building is 50 feet high.

To Solve a Proportion for an Unknown Term	1. Cross multiply to form an equation. 2. Multiply the side of the equation without the unknown term. 3. Divide both sides of the equation by the number that the unknown term is multiplied by. This cancels the number that the unknown term is multiplied by, giving you the value of the unknown term.

Exercises

Find the missing term in each of the following proportions.

1. $\dfrac{4}{18} = \dfrac{10}{N}$

2. $\dfrac{10}{t} = \dfrac{8}{28}$

3. $\dfrac{c}{18} = \dfrac{10}{4}$

4. $\dfrac{4}{1} = \dfrac{52}{S}$

5. $\dfrac{H}{12} = \dfrac{25}{20}$

6. $\dfrac{6}{21} = \dfrac{A}{70}$

7. $\dfrac{12}{40} = \dfrac{F}{9}$

8. $\dfrac{12}{20} = \dfrac{N}{25}$

9. $\dfrac{0.12}{0.28} = \dfrac{s}{42}$

10. $\dfrac{1}{6} = \dfrac{t}{78}$

11. $\dfrac{8}{12} = \dfrac{z}{54}$

12. $\dfrac{1}{17} = \dfrac{c}{51}$

13. $\dfrac{102}{17} = \dfrac{36}{D}$

14. $\dfrac{4}{G} = \dfrac{92}{23}$

15. $\dfrac{K}{40} = \dfrac{14}{16}$

16. $\dfrac{w}{0.16} = \dfrac{0.15}{0.40}$

17. The Seaport Plaza Hotel employs 2 part-time workers for every 1 full-time worker during the summer, and 1 part-time worker for every 2 full-time workers in the winter. The hotel always has 56 full-time workers. How many part-time workers does it have in the summer?

18. Use the information given in Problem 17. How many part-time workers does the hotel have in the winter?

Word Problems

The following problems will give you more practice in using ratios and proportions.

1. The state of Utah has a population of 1,613,404 people. This is about 19 people per square mile. About how many square miles are in Utah?

2. There are 30 inches in 76.2 centimeters. How many inches are in 2.54 centimeters?

3. Last year, 23,999 million pounds of beef were eaten in the United States. If the population was 233 million, how many pounds did each person eat?

4. A 5-foot-tall student compared herself to the Statue of Liberty, 150 feet high. She figured that if she had the same proportions, and her waist was about 1 foot wide, the statue had a waist span of 30 feet. Was she right?

5. Harold bought a 20-ounce box of breakfast cereal for $2.50, saying that he would not have saved any money by buying an 80-ounce box for $10.00. He said that the cost per ounce was the same for both boxes. Was he right?

6. If there are 28 grams in one ounce, how many grams are in 1 pound (16 ounces)?

7. A veterinary assistant read the label on the bottle of penicillin to be given to a 250-pound calf. The directions said to give 8 cc's for every 1000 pounds of animal weight. How many cc's should the calf get?

8. If 3 eggs are added to 2 pounds of flour to make a cake mix, how many eggs must be added to 600 pounds of flour?

Answers

6. 448 grams   7. 2 cc's   8. 900 eggs

1. 84,916 square miles   2. 1 inch   3. 103 pounds   4. Yes   5. Yes

Solving Percent Problems

In the last unit you learned that percent was the number of parts per hundred.

If a survey reports that 46 out of 100 people watch television news daily, then you know that 46% of the people watch the news.

How would you figure the percent of newswatchers where the total in the survey was 60, and 33 of them watched the news?

You know that there are 33 newswatchers in 60 people. To find percent, you want the number of newswatchers per 100 people. This is just like the proportion problems you have been working.

One ratio is 33 to 60. The other ratio is a number that is to 100 as 33 is to 60. You can call this number r, or any other letter.

$$\frac{33}{60} = \frac{r}{100}$$

You solve this proportion in the same manner that you solve any proportion.

$$\frac{33}{60} = \frac{r}{100}$$

$$33 \times 100 = 60 \times r$$

$$3300 = 60 \times r$$

$$\frac{3300}{60} = \frac{60 \times r}{60}$$

$$55 = r$$

Substitute 55 for r in the proportion.

$$\frac{33}{60} = \frac{55}{100}$$

$$33 \times 100 = 60 \times 55$$

$$3300 = 3300$$

Cross multiplication shows that the answer is correct.

$\frac{55}{100} = 55\%$, so 55% of the people surveyed watch the news.

You'll learn more about how to work percent problems in the next unit. First finish this unit by working the review exercises and taking the test.

Review Exercises

Write each ratio in fraction form and reduce to lowest terms.

1. 20 to 2　　　　**2.** 46 to 16　　　　**3.** 8 to 800　　　　**4.** 4 to 400

5. In 100 milliliters of healthy human blood there are about 55 milliliters of plasma and about 45 milliliters of formed elements (red and white blood cells). What is the ratio of plasma to formed elements? Reduce to lowest terms.

Do these pairs of ratios make proportions?

6. $\dfrac{9}{8}$　$\dfrac{13.5}{12}$　　**7.** $\dfrac{1}{14}$　$\dfrac{2}{7}$　　**8.** $\dfrac{3}{10}$　$\dfrac{4.5}{15}$　　**9.** $\dfrac{8}{16}$　$\dfrac{21}{40}$

Find the unknown term in each of the following proportions.

10. $\dfrac{20}{25} = \dfrac{D}{45}$　　**11.** $\dfrac{2}{3} = \dfrac{B}{42}$　　**12.** $\dfrac{2.4}{2.8}$　$\dfrac{18}{A}$　　**13.** $\dfrac{15}{G} = \dfrac{6}{10}$

14. A map is drawn to the scale of 1 inch to 50 miles. What is the actual distance of 3.5 inches on the map?

15. A paper mill claims that for every 50 acres of trees they harvest, 39 acres are reforested. What percent of the harvested land is reforested? $\left(Hint: \dfrac{39}{50} = \dfrac{r}{100}.\right)$

16. A parts supply company had to lay off employees in both of its branch warehouses. The larger warehouse laid off 79 of its 316 employees, and the smaller warehouse laid off 8 of its 32 employees. Set up a ratio to see if the two warehouses were treated fairly.

Answers

1. $\dfrac{1}{10}$　　**2.** $\dfrac{23}{8}$　　**3.** $\dfrac{1}{100}$　　**4.** $\dfrac{1}{100}$　　**5.** $\dfrac{11}{9}$

6. Yes　　**7.** No　　**8.** Yes　　**9.** No　　**10.** $D = 36$

11. $B = 28$　　**12.** $A = 21$　　**13.** $G = 25$　　**14.** 175 miles　　**15.** 78%

16. Yes, the two warehouses were treated fairly.

Unit 10 Test

Use a separate piece of paper to work out each problem.

Do these ratios make proportions?

1. $\dfrac{7}{2}$ $\dfrac{21}{6}$

2. $\dfrac{8}{3}$ $\dfrac{5}{2}$

3. $\dfrac{25}{125}$ $\dfrac{4}{20}$

4. $\dfrac{0.2}{2}$ $\dfrac{0.3}{3}$

5. $\dfrac{0.7}{8}$ $\dfrac{0.8}{8}$

6. $\dfrac{160}{60}$ $\dfrac{24}{9}$

Find the unknown term in each of these proportions.

7. $\dfrac{0.7}{0.9} = \dfrac{K}{36}$

8. $\dfrac{12}{40} = \dfrac{F}{9}$

9. $\dfrac{K}{40} = \dfrac{14}{16}$

10. $\dfrac{9}{B} = \dfrac{57}{19}$

11. $\dfrac{20}{25} = \dfrac{d}{45}$

12. $\dfrac{b}{42} = \dfrac{2}{3}$

13. $\dfrac{2.4}{2.8} = \dfrac{18}{a}$

14. $\dfrac{15}{g} = \dfrac{6}{10}$

15. $\dfrac{k}{36} = \dfrac{0.7}{0.9}$

16. $\dfrac{0.4}{0.9} = \dfrac{4.8}{e}$

17. $\dfrac{13}{26} = \dfrac{5}{r}$

18. On a set of blueprints, 1 inch = 18 feet. How many feet are represented by 4 inches?

19. If a magnifying glass makes something that is 2 millimeters long appear 9 millimeters long, how long will something that is actually 3 millimeters long appear?

20. At a stockholders' meeting 20 out of 80 stockholders attended. These 20 people represented $4,000,000. How much money do all 80 stockholders represent?

Your Answers

1. _____
2. _____
3. _____
4. _____
5. _____
6. _____
7. _____
8. _____
9. _____
10. _____
11. _____
12. _____
13. _____
14. _____
15. _____
16. _____
17. _____
18. _____
19. _____
20. _____

11 Using Percent

In this unit you will work exercises like these:

- What is 45% of 89?

- 63% of what number is 5?

- What percent of 110 is 6?

- What is the yearly amount of interest on $6,000 at a yearly rate of 14%?

Percent Proportion Formula

Every percent statement has three components. It is important to understand what these components are and to be able to identify them. The three components are:

RATE The RATE is always the number with a % sign or the word "percent."
(r or R) (When you see a small r, it means the number by itself. The % sign doesn't go with it. When you see a capital R, it means that the % sign goes with the number. For now, we'll just be using the small r.)

BASE (B) The BASE may be thought of as the "starting point" or the "whole thing."

PART (P) The PART is the result of taking a percent of the base.

These components relate to each other according to the *percent proportion formula:*

$$\frac{\text{Part}}{\text{Base}} = \frac{\text{rate}}{100} \quad \text{or} \quad \frac{P}{B} = \frac{r}{100}.$$

Note that you use a small r for rate, meaning a number *without* a % sign. (The 100 in the proportion takes care of the % sign.)

The percent proportion formula is so important that we don't ever want you to forget it.

$$\frac{P}{B} = \frac{r}{100}$$

Look at this percent statement. The rate, the Base, and the Part have been labeled.

25% of 56 is 14.

$r \qquad B \qquad P$

Here, rate, Base, and Part have been put into the percent proportion.

$$\frac{P}{B} = \frac{r}{100}$$

$$\frac{14}{56} = \frac{25}{100}$$

This proportion means that 14 is to 56 as 25 is to 100. That's what percent is all about! Percent means comparing everything to 100.

Thinking of percent statements as proportions will help you to understand the meaning of percent as well as enable you to solve all types of percent problems.

The percent proportion can be used to solve any percent problem. The only thing to be careful about is identifying the three different components.

Look at this percent statement. The three components of the percent proportion, rate, Base, and Part, have been labeled.

40% of 80 is 32.

r   B   P

The easiest to identify is rate. Just look for the % sign or the word "percent." The rate is the number by itself. The rate *(r)* is 40.

The next easiest to identify is Base. The Base usually comes after the word "of" in a percent problem. Remember that Base represents the whole quantity. The Base *(B)* is 80.

After you have identified rate and Base, Part is the number that's left. The Part is usually found with the word "is" in a percent problem. The part *(P)* is 32.

Working with percentages is easier when you remember what words *r, B,* and *P* each hang around with in a percent problem.

Identifying r, B, and P

So far each exercise set has dealt with a separate type of percent problem. This has made it fairly easy to identify rate *(r)*, Base *(B)*, and Part *(P)*.

Accurately identifying *r, B,* and *P* becomes more difficult and more important when the three different types of percent problems are all mixed together. The next two exercise sets have the three types mixed together, so it's time to sharpen your skills on identifying *r, B,* and *P*.

Example   Identify rate, Base, and Part in this question: What is 23% of 90?

Start with rate; rate is easy. It's always the number followed by the % sign or "percent."

rate *(r)* is 23.

Next look for Base. Base comes after the word "of."

Base *(B)* is 90.

Part is the only number that's left. Here Part is the unknown.

Part *(P)* is unknown.

Another Example   Identify *r, B,* and *P* in the following exercise.

A collection agency collected $563 for a client, charging $67.86 for this service. What was the rate of commission charged?

Start with *r*. There is no % sign or "percent" in this exercise. But the question is "What was the *rate?*". This tells you that *r* is unknown.

r is unknown.

Next look for Base. No "of" in this exercise, so look for the number that represents the "whole thing." $563 is the *whole amount* collected.

B is $563.

Part is the only number that's left. Here Part is the amount the collection agency charged for its service. $67.86 is *part* of $563.

P is $67.86.

Another Example

Identify *r*, *B*, and *P* in this exercise.

In a recent community election 12,467 people voted. This was 38% of the eligible voters. How many eligible voters are there in the community?

12,467 is 38% of what?

Start with *r*.

r is 38.

Next look for Base. Base is the "whole thing." Here Base is the *total number* of voters in the community, which is what you are asked to find.

B is unknown.

Part is the only number that's left. Here Part is the number of people who actually voted. 12,467 people are *part* of the total number of eligible voters.

P is 12,467.

Another Example

Identify *r*, *B*, and *P* in this exercise.

A medical biologist used drug-treated mice to research brain tumors. She expected 13 mice to develop tumors. If 16 mice actually developed tumors, what percent of the expected number is this?

16 is what percent of 13?

Start with *r*. The question asks "what percent," so you know that *r* is unknown.

r is unknown.

Next look for Base. Careful. You want the percent of the *expected number,* so the expected number is the "whole thing."

B is 13.

Part is the only number that's left. Here Part is the actual experimental result.

P is 16.

Now work some exercises on identifying *r*, *B*, and *P*.

Exercises

In each of the following, identify *r, B,* and *P.* Do not solve these problems.

1. 20% of 7 is what number?

2. What number is 78% of 109?

3. What percent of 50 is 30?

4. 80 is 25% of what number?

5. What number is 90% of 3,345?

6. 54 is what percent of 20?

7. What percent of 108 is 98?

8. 45% of what number is 36?

9. 230% of what number is 870?

10. 475% of 23 is what number?

11. Delta Carpet Company loses 5.3% of its carpet in cutting and installation. Of the total 3,078 yards used in one month, how much was lost?

12. A salesman must collect 5% sales tax on his sales. His total sales in January were $11,500. What is the amount of tax he must collect?

13. What amount of savings would be needed to generate $54.20 interest if the rate of return is $4\frac{1}{2}$%?

14. Dan Wright estimates that total sales at his Seven-Eleven for the next month will be $175,150 and that advertising expenditures will amount to $5,200. What percent of the total sales will be spent on advertising?

15. This month a saleswoman at Riegel's Furniture earned $987 on a straight commission of 28% of sales. What were her total sales for the month?

16. In Ontario, the loss of 3 points on a driver's license amounts to 25% of the total points allotted to a driver. How many points are allotted to each driver?

Answers

1. $P = ?, B = 7, r = 20$ 2. $P = ?, B = 109, r = 78$ 3. $P = 30, B = 50, r = ?$

4. $P = 80, B = ?, r = 25$ 5. $P = ?, B = 3,345, r = 90$ 6. $P = 54, B = 20, r = ?$

7. $P = 98, B = 108, r = ?$ 8. $P = 36, B = ?, r = 45$ 9. $P = 870, B = ?, r = 230$

10. $P = ?, B = 23, r = 475$ 11. $P =$ the amount lost, $B = 3,078, r = 5.3$

12. $P =$ the amount of tax, $B = \$11,500, r = 5$

13. $P = \$54.20, B =$ the amount of savings, $r = 4\frac{1}{2}$

14. $P = \$5,200, B = \$175,150, r = ?$ 15. $P = \$987, B =$ total sales, $r = 28$

16. $P = 3$ points, $B =$ total points, $r = 25\%$

Finding Part

Look again at the percent proportion formula. $\dfrac{P}{B} = \dfrac{r}{100}$

There are only three unknown terms, r, B, and P. Therefore, there are only three different types of percent problems.

The first type of problem tells you the values of r and B, but not of P. In this type of problem you must solve for P.

Example   What is 23% of 90?

$\quad\quad P \quad\quad r \quad\quad B$ First identify rate, Base, and Part
($r \rightarrow$ %, $B \rightarrow$ "of," $P \rightarrow$ "is").

$\dfrac{P}{B} = \dfrac{r}{100}$ Now put the values of r and B into the percent proportion formula.

$\dfrac{P}{90} = \dfrac{23}{100}$ This formula is just like the proportions from the last unit.

$P \times 100 = 90 \times 23$ Solve for P.

$P \times 100 = 2070$

$\dfrac{P \times 100}{100} = \dfrac{2070}{100}$

$P = 20.7$ The answer is 20.7.
20.7 is 23% of 90.

Another Example   116% of 50 is what number?

$\quad\quad r \quad\quad B \quad\quad\quad P$ Identify rate, Base, and Part.

$\dfrac{P}{B} = \dfrac{r}{100}$ Set up a proportion and solve for P.

$\dfrac{P}{50} = \dfrac{116}{100}$

$P \times 100 = 50 \times 116$

$P \times 100 = 5800$

$\dfrac{P \times 100}{100} = \dfrac{5800}{100}$

$P = 58$ The answer is 58.
116% of 50 is 58.

Don't worry if you're still confused about how to identify r, B, and P. It will become easier as you work various types of percent problems.

Exercises

Solve these exercises by identifying *r, B,* and *P,* and setting up a proportion to solve for *P.*

1. What is 15% of 36?

2. 6% of 600 is what number?

3. 15% of 200 is what number?

4. What number is 20% of 400?

5. What is 150% of 14?

6. 7.9% of 100 is what number?

In the following problems, round off each answer to the nearest hundredth.

7. What is 0.3% of 126?

8. 30% of 1200 is what number?

9. What is 2.5% of 750?

10. What is $2\frac{1}{2}$% of 7?

11. What is 0.4% of 60?

12. 6% of 3.1 is what number?

13. A retailer has $23,000 invested in his business and finds that he is earning 12% per year on his investment. How much money is he making per year?

14. The widow of a bank president recently purchased a Florida condominium for $82,000. If she expects a return on her investment of 8% a year, what will this amount to in dollars per year?

Answers

14. 8% of $82,000 is $6,560.

13. 12% of $23,000 is $2,760.

1. 5.4 **2.** 36 **3.** 30 **4.** 80 **5.** 21 **6.** 7.9 **7.** 0.38 **8.** 360 **9.** 18.75 **10.** 0.18 **11.** 0.24 **12.** 0.19

Finding Base

Ready for a new type of percent problem?

In the second type of percent problem you know the values of *r* and *P*, but not of *B*. You still need the percent proportion formula.

$$\frac{P}{B} = \frac{r}{100}$$

You use the percent proportion formula to set up a proportion and solve for *B*.

Example 843 is 25% of what number?

P	*r*		*B*

Identify rate, Base, and Part.

$$\frac{P}{B} = \frac{r}{100}$$

Put the value of *r* and *P* into the percent proportion formula.

$$\frac{843}{B} = \frac{25}{100}$$

Solve for *B*.

$$843 \times 100 = B \times 25$$

$$84300 = B \times 25$$

$$\frac{84300}{25} = \frac{B \times 25}{25}$$

$$\frac{84300}{25} = B$$

$$3{,}372 = B$$

The answer is 3,372.
843 is 25% of 3,372.

Another Example 80% of what number is 68?

r	*B*	*P*

Identify rate, Base, and Part.

$$\frac{P}{B} = \frac{r}{100}$$

Set up a proportion and solve for *B*.

$$\frac{68}{B} = \frac{80}{100}$$

$$68 \times 100 = B \times 80$$

$$6800 = B \times 80$$

$$\frac{6800}{80} = \frac{B \times 80}{80}$$

$$85 = B$$

The answer is 85.
80% of 85 is 68.

Exercises on finding Base are next.

Exercises

Solve these exercises by identifying *r, B,* and *P* and setting up a proportion to solve for *B.*

1. 3 is 6% of what number?

2. 25% of what number is 16?

3. 12 is 6% of what number?

4. 30 is 25% of what number?

5. 4% of what number is 2?

6. 10% of what number is 18?

7. 46 is 200% of what number?

8. 14 is 20% of what number?

9. 9 is 20% of what number?

10. 19 is 95% of what number?

11. Sharon McNett saves 23% of her salary toward the down payment on a larger home. If this amounts to $276 per month, find her monthly salary?

12. A warehouse supervisor owns common stock in his company valued at $3,078, which amounts to 17% of his total investments. What is the value of his total investments?

Answers

12. $3,078 is 17% of $18,105.88.
11. $276 is 23% of $1,200.

10. 20   **9.** 45   **8.** 70   **7.** 23   **6.** 180
5. 50   **4.** 120   **3.** 200   **2.** 64   **1.** 50

Finding Rate

In the third and last type of percent problem you know the values of B and P, but not of r. You use the percent proportion formula to solve for r.

$$\frac{P}{B} = \frac{r}{100}$$

Be careful when you identify B and P. They are easily mixed up. Remember that B represents the "whole thing" and usually comes after the word "of."

Example   What percent of 68 is 4? Round off to the nearest hundredth.

r	B	P	Identify rate, Base, and Part.

$$\frac{P}{B} = \frac{r}{100}$$
Put the values of B and P into the percent proportion formula.

$$\frac{4}{68} = \frac{r}{100}$$
Solve for r.

$$4 \times 100 = 68 \times r$$

$$400 = 68 \times r$$

$$\frac{400}{68} = \frac{68 \times r}{68} \qquad 400 \div 68 = 5.88$$

$$5.88 = r$$
The answer is 5.88%.
5.88% of 68 is 4.

Note that r is just a number. The question asks for percent, so you must add a % sign.

Another Example   14 is what percent of 3? Express the percent as a mixed number.

P	r	B	Identify rate, Base, and Part.

$$\frac{P}{B} = \frac{r}{100}$$
Set up a proportion and solve for r.

$$\frac{14}{3} = \frac{r}{100}$$

$$14 \times 100 = 3 \times r$$

$$1400 = 3 \times r$$

$$\frac{1400}{3} = \frac{3 \times r}{3} \qquad 1400 \div 3 = 466\frac{2}{3}$$

$$466\frac{2}{3} = r$$
The answer is $466\frac{2}{3}$%.
14 is $466\frac{2}{3}$% of 3.

Exercises

Solve these exercises by identifying *r, B,* and *P* and setting up a proportion to solve for *r*.

1. 36 is what percent of 9?

2. What percent of 60 is 12?

3. 27 is what percent of 300?

4. 50 is what percent of 40?

5. What percent of 50 is 1?

6. 13 is what percent of 50?

For each of the following, express the percent as a mixed number.

7. 3 is what percent of 24?

8. 5 is what percent of 12?

9. What percent of 7 is 4?

10. 32 is what percent of 36?

11. What percent of a mail order firm's merchandise is damaged if total shipments are $278,000 and damage is equal to $3,840? (Round off to one decimal place.)

12. If the cost of a new car is $205 more than last year's price of $8,500, what is the percent of increase? (Round off to one decimal place.)

Answers

1. 400% **2.** 20% **3.** 9% **4.** 125% **5.** 2% **6.** 26% **7.** $12\frac{1}{2}$% **8.** $41\frac{2}{3}$% **9.** $57\frac{1}{7}$% **10.** $88\frac{8}{9}$% **11.** 1.4% of $278,000 is $3,840. **12.** $2.4\overline{4}$% of $8,500 is $205.

Combined Percent Exercises

Below are the three types of percent problems in mixed order. Be sure to identify r, B, and P properly before you set up a proportion. Then solve the problems.

1. 10 is 20% of what number?

2. 34 is what percent of 136?

3. 60 is what percent of 200?

4. 14 is 25% of what number?

5. 24% of 200 is what number?

6. 6 is what percent of 4?

7. 18 is 25% of what number?

8. 30% of what number is 9?

9. 150% of 8 is what number?

10. 7% of what number is 14?

11. Sid's pharmacy has a total monthly payroll of $6,875, of which $1,450 goes toward employee fringe benefits. What percent of the total payroll is used for fringe benefits? (Round off to the nearest decimal place.)

12. In analyzing the success of license applicants, the state finds that 58.3% of those examined received a passing mark. The records show that 8,370 new licenses were issued. What was the number of applicants? (Round off to the nearest whole number.)

13. Senator Savings Bank pay $5\frac{1}{4}$% interest per year. What is the annual interest on an account of $830?

14. A survey of 300 drivers showed that only 63% wore their seat belts. How many of the drivers wore seat belts?

Answers

11. 21.1%	12. 14,357 applicants	13. $43.58	14. 189 drivers	
6. 150%	7. 72	8. 30	9. 12	10. 200
1. 50	2. 25%	3. 30%	4. 56	5. 48

Another Way of Solving Percent Problems (Optional)

Now that you know how to solve percent problems using proportions, you are ready to learn another way to solve percent problems.

This doesn't mean that you should forget proportions! Proportions are perhaps the best way to understand the idea of percent. Once the idea is down pat, however, you may like to use this second method to solve percent problems. If you have problems with this new method, go back to using proportions.

Back in the beginning of this unit, you first learned about the three components of a percent problem: rate, Base, and Part.

As you know, the components relate to each other like this:

$$\frac{\text{Part}}{\text{Base}} = \frac{\text{rate}}{100} \quad \text{or} \quad \frac{P}{B} = \frac{r}{100}.$$

In the percent proportion formula you use a small r to mean a number without the percent sign. (The 100 in the percent proportion takes care of the % sign.) If you use a number *with* the % sign (shown by a capital R) you don't need to divide by 100.

So a capital R can replace $\frac{r}{100}$ and change the percent proportion into this formula:

$$\frac{\text{Part}}{\text{Base}} = \text{Rate} \quad \text{or} \quad \frac{P}{B} = R.$$

You usually see this formula flipped around.

$$\text{Rate} = \frac{\text{Part}}{\text{Base}} \quad \text{or} \quad R = \frac{P}{B}$$

This formula can be rearranged to make two other formulas.

$$\text{Base} = \frac{\text{Part}}{\text{Rate}} \quad \text{or} \quad B = \frac{P}{R}$$

$$\text{Part} = \text{Rate} \times \text{Base} \quad \text{or} \quad P = R \times B$$

These three formulas can be used to solve the three types of percent problems.

To Solve Percent Problems Using Formulas	Identify the components. Choose the correct formula. Put the values from the problem into the formula and solve.

Example   What is 34% of 70?

P	R	B	First identify Rate, Base, and Part.

$P = R \times B$ — P is unknown, so use the formula for P.

$P = 34\% \times 70$ — Put the values for R and B into the formula. (R is 34%, not just 34.)

$P = 0.34 \times 70$ — You must change 34% to a decimal before you multiply.

$P = 23.8$ — The answer is 23.8.
23.8 is 34% of 70.

Note that before you can multiply or divide a percent, you must change it to a decimal or a fraction. Just change it to whichever makes the problem easier.

Another Example $66\frac{2}{3}\%$ of what number is 60?

R	B	P	Identify Rate, Base, and Part.

$B = \dfrac{P}{R}$ — B is unknown, so use the formula for B.

$B = \dfrac{60}{66\frac{2}{3}\%} = 60 \div 66\frac{2}{3}\%$ — Put the values of R and P into the formula.

$66\frac{2}{3}\% = 66\frac{2}{3} \div \dfrac{100}{1}$ — Change $66\frac{2}{3}\%$ to a fraction.

$\dfrac{200}{3} \times \dfrac{1}{100} = \dfrac{200}{300}$ or $\dfrac{2}{3}$

$B = 60 \div \dfrac{2}{3} = 60 \times \dfrac{3}{2}$ — Put $\frac{2}{3}$ where $66\frac{2}{3}\%$ was before.

$B = 90$ — The answer is 90. $66\frac{2}{3}\%$ of 90 is 60.

Another Example   13 is what percent of 20?

P	R	B	Identify Rate, Base, and Part.

$R = \dfrac{P}{B}$ — R is unknown, so use the formula for R.

$R = \dfrac{13}{20}$ — Put the values of B and P into the formula.

$R = 0.65$ — Write 0.65 as a percent.

$R = 65\%$ — The answer is 65%.
13 is 65% of 20.

Note that when you need to find R, your answer will first be a fraction or a decimal. You then write R in percent form.

Now try some exercises on using formulas to solve percent problems.

Exercises

For each of the following, use this formula to find the value of P: $P = R \times B$.

1. What number is 25% of 36?

2. 5% of 1205 is what number?

3. What number is 45% of 10?

4. $12\frac{1}{2}$% of 48 is what number?

For each of the following, use this formula to find the value of B: $B = \frac{P}{R}$.

5. 6 is 6% of what number?

6. 12% of what number is 63?

7. $133\frac{1}{3}$% of what number is 16?

8. 45 is 250% of what number?

For each of the following, use this formula to find the value of R: $R = \frac{P}{B}$. If the percent is not a whole number, write it as a mixed number.

9. What percent of 64 is 16?

10. What percent of 70 is 7?

11. 0.7 is what percent of 35?

12. What percent of 63 is 9?

Answers

9. 25%	10. 10%	11. 2%	12. $14\frac{2}{7}$%
5. 100	6. 525	7. 12	8. 18
1. 9	2. 60.25	3. 4.5	4. 6

Exercises

Work each of the following percent problems using the formula method. The three types are mixed together, so be sure to use the correct formula. Express a percent that is not a whole number as a mixed number. Round off other numbers to the nearest hundredth if needed.

1. $33\frac{1}{3}\%$ of 27 is what number?

2. What number is 24% of 46?

3. 56 is what percent of 128?

4. 13.3 is what percent of 19?

5. A home valued at $58,000 today is said to be gaining value at the rate of 12% per year. Find the value of the home one year from now.

6. This month's sales goals for Big Pen Company is 2,380,000 ball-point pens. If sales of 1,844,500 pens have been made, what percent of the goal remains?

7. On sales over $100 at Beth's Bargain Basement, a discount of 15% is allowed on the entire amount of the sale. What would the discount be on a sale of $135?

8. For a certain theater, entertainment taxes amount to 10% of the ticket prices. If the amount collected in taxes is $352, what is the amount of money received from ticket sales?

Answers

1. 9   2. 11.04   3. $43\frac{3}{4}\%$ 4. 70%

5. 12% of $58,000 is $6,960. $58,000 + $6,960 = $64,960

6. $22\frac{1}{2}\%$ 7. $20.25   8. $3,520

Interest Problems

One common use of percents is in the calculation of interest problems. When you borrow money, you pay interest for using the money. The interest is a percent of the amount you borrow.

When you save money in a savings account, the bank or savings and loan pays you interest. The interest is a percent of the amount you have in the account.

Example   What is the yearly interest on $3,500 when the yearly rate is 8%?

You want to know what 8% of $3,500 is. You can either work the problem using a proportion or using the proper formula. Here is the proportion method.

$$\frac{P}{B} = \frac{r}{100}$$

$$\frac{P}{3,500} = \frac{8}{100}$$

$$P \times 100 = 3,500 \times 8$$

$$P \times 100 = 28,000$$

$$P = 280$$

The interest for one year is $280.

Another   If you borrow $500 at 12% yearly interest, and pay it back at the end
Example   of one year, how much will you have to pay?

First find the amount of interest you will owe.

$$\frac{P}{B} = \frac{r}{100}$$

$$\frac{P}{500} = \frac{12}{100}$$

$$P \times 100 = 500 \times 12$$

$$P \times 100 = 6,000$$

$$P = 60$$

Remember you have to pay back the amount you borrowed as well as the interest.

$$\begin{array}{r} \$500 \\ + \$\ \ 60 \\ \hline \$560 \end{array}$$

At the end of one year you owe $560.

Exercises

1. If you put $850 in a savings account that has 5% yearly interest, how much money do you have at the end of the year?

2. If you borrow $850 at a yearly interest rate of 11%, how much do you owe at the end of one year?

3. If you borrow $400 at a yearly interest rate of 14% and pay back $200 after one year, how much do you owe? (*Hint:* find the amount you owe after one year and subtract $200.)

4. If you put $16,000 in a savings account that has $7\frac{1}{2}$% yearly interest, how much money do you have after one year?

5. If you borrow $24 at an interest rate of 3% per month, how much do you owe at the end of one month?

6. If you put $455 in a savings account that gives an interest rate of 6% yearly, will you have enough money after one year to buy a motorcycle for $480?

7. If you put $896 in a savings account with a yearly interest of $4\frac{1}{2}$%, and take $375 out after one year, how much money will you have?

8. Jack borrowed $900 from Honest Ernie to pay for his college tuition and books. His interest was $16\frac{1}{2}$% per year. Gerri borrowed the same amount from her aunt at only $3\frac{1}{2}$% interest. How much more did Jack have to pay in interest?

Answers

1. $892.50   2. $943.50   3. $256   4. $17,200   5. $24.72
6. Yes   7. $561.32   8. $148.50 (Jack) − $31.50 (Gerri) = $117.00

Sales Tax

Most areas now have a sales tax, so this is a type of percent problem that we have to deal with almost every day. Here are some examples of the kinds of percent problems you are likely to meet.

Example John Crotty bought a new organ for $2,875.25. If the sales tax is 7%, how much did he pay altogether?

$$7\% \text{ of } \$2,875.25 \text{ is how much?}$$

$$r \qquad\qquad B \qquad\qquad P$$

$$\frac{P}{B} = \frac{r}{100}$$

$$\frac{P}{2875.25} = \frac{7}{100}$$

$$P \times 100 = 2875.25 \times 7$$

$$P \times 100 = 20126.75$$

$$P = \$201.27$$

The organ cost $2,875.25 + $201.27 in sales tax, or $3,076.52 altogether.

Example Donna Greenfield paid 21 cents sales tax ($0.21) on a bag of kitty litter worth $3.00. How much was the rate of tax?

$$\$0.21 \text{ is what percent of } \$3.00?$$

$$P \qquad\qquad r \qquad\qquad B$$

$$\frac{P}{B} = \frac{r}{100}$$

$$\frac{0.21}{3.00} = \frac{r}{100}$$

$$3.00 \times r = 0.21 \times 100$$

$$3.00 \times r = 21.00$$

$$r = 7\%$$

The rate of sales tax was 7%.

Example Joe Caldwell paid $1.32 sales tax on his public utilities bill. If the rate of sales tax is 4%, how much was his total bill?

$1.32 is 4% of what?

$$P \qquad r \qquad B$$

$$\frac{P}{B} = \frac{r}{100}$$

$$\frac{1.32}{B} = \frac{4}{100}$$

$$4 \times B = 1.32 \times 100 = 132$$

$$B = 33$$

The total bill is $33.00.

Exercises

Here are some word problems to give you some practice in working with sales tax.

1. Marion mailed a check for $28.50 for two new mail-order lawn chairs. The company sent back her check because she forgot to include $1.71 for the sales tax. What was the rate of sales tax?

2. In Ontario there is no sales tax on food. However, an item such as peanut butter is considered a luxury, and 7% sales tax is charged. How much tax is there on a jar of peanut butter priced at $3.75? (Round off your answer to the nearest cent.)

3. Sales tax rates vary in different areas. In Kentucky, a meal worth $11.25 costs $11.81 with tax included. What is the rate of sales tax on food in Kentucky? (Round off your answer to the nearest percent.)

4. In Michigan, 4% sales tax is charged for motel lodging. If a traveler paid $2.65 tax, what was the cost of the room?

5. Suppose you are working as a sales clerk, and you have to charge 6% sales tax for an item that costs $4.98. What should you ask the customer to pay altogether? (Round off your answer to the nearest cent.)

6. Frank paid $726.15 sales tax for a new car. If the tax rate was 3%, what was the value of the car?

Answers

1. 6% 2. $0.26 or 26 cents 3. 5% 4. $66.25 5. $5.28 6. $24,205.00

Review Exercises

Solve each of the following percent problems. Work them whichever way you choose, proportion or formula. Express percents that are not whole numbers as mixed numbers.

1. 88% of 50 is what number?

2. 2.4 is what percent of 30?

3. 31 is 25% of what number?

4. 3 is $7\frac{1}{2}$% of what number?

5. 98% of what number is 147?

6. What number is $33\frac{1}{3}$% of 90?

7. $37\frac{1}{2}$% of what number is 45?

8. 10 is what percent of 75?

9. What number is 94% of 50?

10. 22 is what percent of 132?

Answers

1. 44	**2.** 8%	**3.** 124	**4.** 40	**5.** 150
6. 30	**7.** 120	**8.** $13\frac{1}{3}$%	**9.** 47	**10.** $16\frac{2}{3}$%

11. 3 is 25% of what number?

12. What percent of 86 is 30?

13. 16 is what percent of 400?

14. What number is 250% of 40?

15. An ad for steel-belted radial tires promises 15% better mileage. If mileage has been 17.5 miles per gallon in the past, what mileage could be expected after installing the new tires? (Round off to the nearest hundredth of a mile.)

16. Helen Scott received $480 annual interest on an investment that paid 4% interest per year. How much did she have invested?

17. Steve Jonston spends 22% of his income on housing, 24% on food, 8% on clothing, 15% on transportation, 11% on education, 7% on recreation, and saves the rest. If his savings amount to $62 per month, what are his monthly earnings? (*Hint:* to find the percent of his income that he saves, add up the percents he spends and subtract from 100%.)

18. Smoke alarms normally priced at $33.90 are reduced by $10.17. Find the percent of reduction.

Answers

11. 12   12. $34\frac{38}{43}$% or 34.88%   13. 4%   14. 100   15. 20.13   16. $12,000   17. $476.92   18. 30%

Unit 11 Test

Use a separate piece of paper to work out each problem.

1. 12 is what percent of 60?

2. 13 is 25% of what number?

3. 38% of 200 is what number?

4. What number is 4% of 140?

5. 16 is 5% of what number?

6. 26 is what percent of 100?

7. 8% of what number is 6?

8. 150% of what number is 75?

9. 21 is 25% of what number?

10. 64 is what percent of 32?

11. What number is 20% of 50?

12. 12 is what percent of 48?

13. $14\frac{3}{4}$% of $750.25 is what?

14. A corporation paid its employees 8% of total sales in the form of profit sharing. If total sales were $275,000, what amount was shared with employees?

15. A building is insured for 80% of its value. If the insurance coverage is $16,000, what is the value of the building?

16. An electric pencil sharpener is marked "reduced 15%, now only $30.09." Find the original price of the pencil sharpener.

17. Judy Lang received her annual credit union dividend statement showing $213.60 in dividends. If her credit union share balance was $3,560 for the year, what annual rate of return was paid by the credit union?

18. If you borrow $670 at a yearly interest rate of $13\frac{1}{2}$%, how much do you owe after one year?

Your Answers

1. _____

2. _____

3. _____

4. _____

5. _____

6. _____

7. _____

8. _____

9. _____

10. _____

11. _____

12. _____

13. _____

14. _____

15. _____

16. _____

17. _____

18. _____

Review Test

Use a separate piece of paper to work out each problem.

1. Write $\frac{3}{16}$ as a decimal and as a percent.

2. Change 0.475 to a percent and a fraction.

3. Change 68% to a decimal and a fraction.

4. Change 2.08 to a percent and a fraction.

5. Write 316% as a fraction and a decimal.

6. Change $2\frac{4}{5}$ to a decimal and a percent.

7. Do the ratios $\frac{10}{1.2}$ and $\frac{1.5}{0.18}$ make a proportion?

8. Is the ratio $\frac{41}{3.8} = \frac{205}{19}$ balanced, or proportional?

9. Is the ratio $\frac{14}{19} = \frac{42}{56}$ proportional?

10. Solve this ratio to find the unknown letter, M: $\frac{M}{5} = \frac{8}{10}$.

11. What is the value of y in $\frac{3}{14} = \frac{y}{49}$?

12. Solve for the unknown in the ratio $\frac{9}{x} = \frac{6}{108}$.

13. What is 128% of 50?

14. 246% of 150 is what number?

15. 86 is what percent of 430?

16. 19% of what number is 9.5?

Solve these word problems.

17. Joe's uncle loaned him $3,250 at 3% interest to buy a car. How much did Joe have to pay his uncle altogether?

18. Scott ate 3 pieces of a pie that had been cut into 8 pieces. What percent of the pie did he eat? What fraction was this?

19. Fifteen out of 90 workers were on vacation the same week. What percent of the workers were away?

20. Six pounds of foot pressure on a car brake pedal create about 50 pounds of braking pressure on the wheels. If the foot pressure is 84 pounds, how much is the pressure on the wheels?

Your Answers

1. _____

2. _____

3. _____

4. _____

5. _____

6. _____

7. _____

8. _____

9. _____

10. _____

11. _____

12. _____

13. _____

14. _____

15. _____

16. _____

17. _____

18. _____

19. _____

20. _____

12 | The Metric System

Aims you toward:

- Understanding the meaning of measurement.
- Understanding the benefits of the metric system.
- Knowing how to use the metric system.
- Knowing how to make simple "guesstimate" conversions from the metric system to the English system.

In this unit you will work exercises like these:

- 45.6 kilometers is how many dekameters?
- 17 milliliters is how many centiliters?

Measurement

When you measure something, all you are doing is comparing it to a particular unit of length, weight, or volume.

The English System

In the ENGLISH SYSTEM, also used in North America, one of the earliest units for length was the distance between the tip of the nose and the tip of the finger on an outstretched hand. This unit, called a yard, was, and still is, used to measure things like the distance across a field.

To measure things like the length of a room people used feet. One foot was the length of a person's foot.

To measure smaller lengths the digit of a thumb was used. This was called an inch.

Similar units of measurement were developed for weight and liquid volume. Grains of barley, for example, were used to weigh small objects.

There are two basic problems with a system of measurement that uses these types of unit measurements.

First, measurement would differ depending upon who was doing the measuring. A short-armed person would measure a shorter yard than a long-armed person.

Some stability was achieved when measurements were based on the King's arm, the King's foot, and so on. But kings do not reign forever.

The problem was finally solved by making a yard, a foot, an inch, and so on independent of human anatomy. Now when a short-armed person measures a yard it is the same length as a yard measured by a long-armed person.

A second problem with the English measurement system is that the relationships between units are all different. There are 12 inches to a foot, but 3 feet to a yard, 1760 yards to a mile, 16 ounces to a pound, 2000 or 2440 pounds to a ton, 8 fluid ounces to a cup, 2 cups to a pint, 2 pints to a quart, and 4 quarts to a gallon. There is a lot of remembering when you use the English system of measurement!

The Metric System and Metric Length

Things are more orderly in the METRIC SYSTEM.

In the METRIC SYSTEM there is one base unit for length, one base unit for weight, and one base unit for liquid volume. To get units of different sizes the base unit is divided or multiplied by 10.

The base unit of *length,* the METER, is divided into 10 parts.

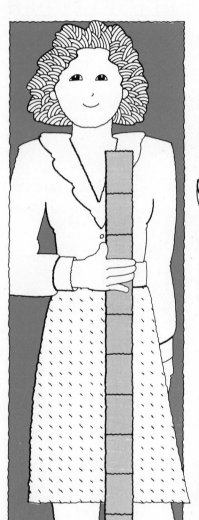

Each part is called a DECIMETER. Each decimeter is divided into 10 parts.

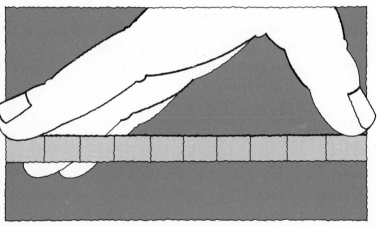

Each of these parts is called a CENTIMETER. Each centimeter is divided into 10 parts.

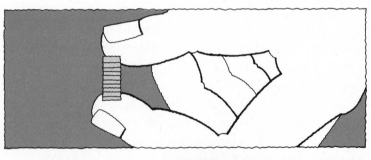

Each part of a centimeter is called a MILLIMETER.

See how easy it is?

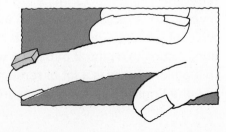

For units to measure longer lengths, the meter is multiplied by 10.

10 meters put together make one DEKAMETER.

10 dekameters put together make one HECTOMETER.
A hectometer is slightly longer than a football field.

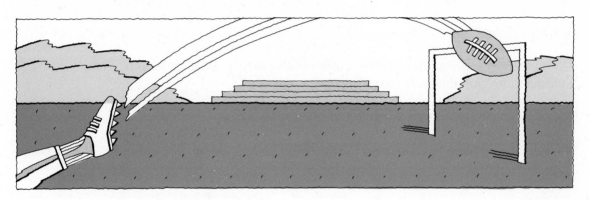

10 hectometers put together make one KILOMETER.
A kilometer is about two-thirds of a mile.

The unit of metric length you choose to use depends upon what you are measuring. If you are measuring long distances, you might use kilometers. If you are measuring something very small, you might use millimeters.

Look at the prefixes for each unit. They tell how each unit relates to the meter.

Kilo	means 1,000.	A kilometer is 1,000 times a meter.
Hecto	means 100.	A hectometer is 100 times a meter.
Deka	means 10.	A dekameter is 10 times a meter.
Deci	means $\frac{1}{10}$.	A decimeter is $\frac{1}{10}$ of a meter.
Centi	means $\frac{1}{100}$.	A centimeter is $\frac{1}{100}$ of a meter.
Milli	means $\frac{1}{1000}$.	A millimeter is $\frac{1}{1000}$ of a meter.

Here is a diagram that shows the relation of the metric units for length. (The abbreviations are shown in parentheses.)

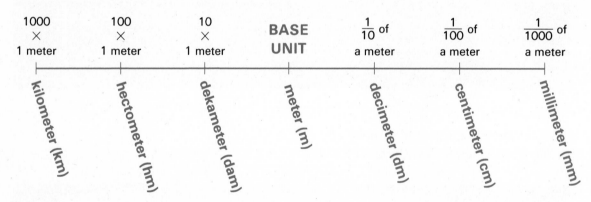

The units used most often are km, m, cm, and mm.

Conversions from one unit of length to another unit of length are much easier than in the English system.

To Change from One Metric Unit to Another	Use the diagram above to count the number of places from the unit you have to the unit you want.
	Move the decimal point the same direction and the same number of places.

Example   1 kilometer is how many meters?

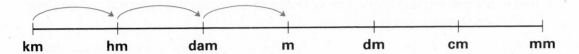

Meters are 3 places to the right of kilometers, so move the decimal point 3 places to the right. (Remember that the unseen decimal point in 1 kilometer goes here: 1. kilometer.)

1000.

1 kilometer = 1000 meters

Another Example   3000 millimeters is how many dekameters?

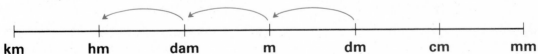

There are 4 places to the left from millimeters to dekameters, so move the decimal point 4 places to the left. (The decimal point starts here: 3000. millimeters.)

0.3000.

3000 millimeters = 0.3 dekameter

Another Example   How many hectometers are there in 14.6 decimeters?

There are 3 steps to the left from decimeters to hectometers, so move the decimal point 3 places to the left.

0.0146

14.6 decimeters = 0.0146 hectometer

Remember when you use the scale that if you move to the left, the decimal point moves to the left. If you move to the right, the decimal point moves to the right.

Now work the exercises on the next page.

Exercises

1. Fill in the blanks below this line with the units for metric length. Then use the line to do the rest of the exercises.

|———————|———————|———————|———————|———————|———————|

__ km __   ____   ____   ____   ____   ____   ____

2. 50 meters = _____ centimeters

3. 85.5 millimeters = _____ meter

4. 0.42 hectometer = _____ decimeters

5. 66 cm = _____ dm

6. 16 km = _____ dam

7. 20 millimeters = _____ centimeters

8. 58 m = _____ dm

9. 1 meter = _____ kilometer

10. 1 meter = _____ millimeters

11. 30 dm = _____ mm

12. 120 centimeters = _____ meters

13. 278 mm = _____ dm

14. 1 mm = _____ km

15. 1 km = _____ mm

16. 30 dekameters = _____ decimeters

17. 140 meters = _____ kilometer

18. 360 cm = _____ m

19. 4 km = _____ dm

Answers

17. 0.140 **18.** 3.6 **19.** 40,000

12. 1.2 **13.** 2.78 **14.** 0.000001 **15.** 1,000,000 **16.** 3000

7. 2 **8.** 580 **9.** 0.001 **10.** 1000 **11.** 3000

2. 5000 **3.** 0.0855 **4.** 420 **5.** 6.6 **6.** 1600

1.

km	hm	dam	m	dm	cm	mm

|———————|———————|———————|———————|———————|———————|

Metric Weight

In the metric system the base unit for *weight* is the GRAM. It takes 454 grams to make a pound.

The gram is divided by 10 to get the units used to measure things that are very light in weight.

The gram is multiplied by 10 to get the units used to measure heavier things.

The prefixes and their meanings are the same as before.

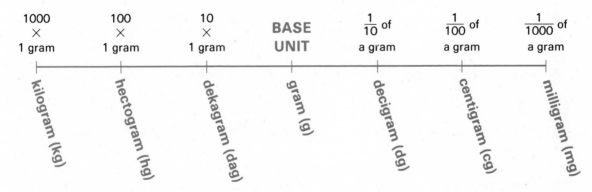

1000 × 1 gram	100 × 1 gram	10 × 1 gram	BASE UNIT	$\frac{1}{10}$ of a gram	$\frac{1}{100}$ of a gram	$\frac{1}{1000}$ of a gram
kilogram (kg)	hectogram (hg)	dekagram (dag)	gram (g)	decigram (dg)	centigram (cg)	milligram (mg)

The units used most often are kg, g, and mg.

To convert from one unit of metric weight to another unit of metric weight, use the same methods you learned for length.

Example   0.078 kilogram is how many grams?

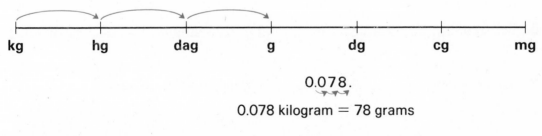

0.078.

0.078 kilogram = 78 grams

Another Example   4,700 centigrams is how many dekagrams?

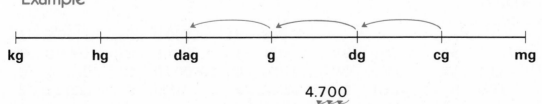

4.700

4,700 centigrams = 4.7 dekagrams

Exercises on metric weight are next.

Exercises

1. Fill in the blanks below this line with the units for metric weight. Then use the line to do the rest of the exercises.

| ├────────────┼────────────┼────────────┼────────────┼────────────┼────────────┤ |

kg _____   _____   _____   _____   _____   _____   _____

2. 14 dag = _____ cg

3. 6 grams = _____ centigrams

4. 312.4 hg = _____ kg

5. 1024 dg = _____ dag

6. 3700 cg = _____ hg

7. 116.3 dag = _____ dg

8. 5 mg = _____ kg

9. 13 cg = _____ dag

10. 48 decigrams = _____ milligrams

11. 0.1 gram = _____ milligrams

12. 0.1 gram = _____ kilogram

13. 900 dekagrams = _____ decigrams

14. 12 hectograms = _____ dekagrams

15. 3.8 dg = _____ mg

16. 127.6 cg = _____ dag

17. 6.11 grams = _____ hectogram

18. 33 decigrams = _____ kilogram

19. 84 cg = _____ mg

Answers

kg	hg	dag	g	dg	cg	mg
├───────┼───────┼───────┼───────┼───────┼───────┤ **1.**

2. 14,000   **3.** 600   **4.** 31.24   **5.** 10.24   **6.** 0.37

7. 11,630   **8.** 0.000005   **9.** 0.013   **10.** 4800   **11.** 100

12. 0.0001   **13.** 90,000   **14.** 120   **15.** 380   **16.** 0.1276

17. 0.0611   **18.** 0.0033   **19.** 840

Metric Liquid Volume

In the metric system the base unit for *liquid volume* is the LITER. A liter is about the same as a quart.

The smaller or larger units for liquid volume are merely the liter divided or multiplied by ten. Again, the prefixes are the same.

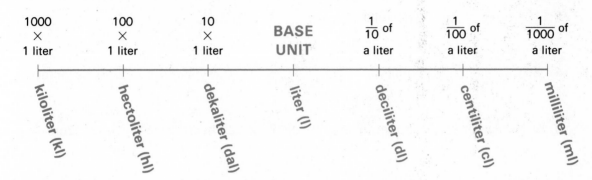

1000 × 1 liter	100 × 1 liter	10 × 1 liter	BASE UNIT	$\frac{1}{10}$ of a liter	$\frac{1}{100}$ of a liter	$\frac{1}{1000}$ of a liter
kiloliter (kl)	hectoliter (hl)	dekaliter (dal)	liter (l)	deciliter (dl)	centiliter (cl)	milliliter (ml)

The units used most often are kl, l, and ml.

Use this chart in the usual way to convert from one unit of liquid volume to another unit of liquid volume.

Example   5.4 liters is how many milliliters?

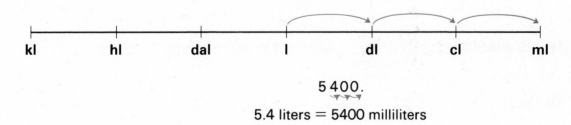

5 4 0 0.

5.4 liters = 5400 milliliters

Another Example   How many dekaliters is 54 centiliters?

0.054

54 centiliters = 0.054 dekaliter

Exercises on metric liquid volume are next.

Exercises

1. Fill in the blanks below this line with the units for metric liquid volume. Then use the line to do the rest of the exercises.

kl	___	___	___	___	___	___

2. 108 liters = _____ hectoliters

3. 42 ml = _____ l

4. 6 liters = _____ centiliters

5. 1862 milliliters = _____ dekaliter

6. 0.04 kiloliter = _____ centiliters

7. 22.22 hectoliters = _____ liters

8. 406.5 dl = _____ hl

9. 48 deciliters = _____ milliliters

10. 0.108 hectoliters = _____ liters

11. 6.2 liters = _____ kiloliter

12. 900 dekaliters = _____ deciliters

13. 387 cl = _____ dal

14. 7,309 dl = _____ ml

15. 0.0839 kl = _____ l

16. 78.406 hl = _____ l

17. 892,644 cl = _____ hl

18. 78 liters = _____ dl

19. 3.1 hectoliters = _____ dekaliters

Answers

1.

kl	hl	dal	l	dl	cl	ml

2. 1.08   **3.** 0.042   **4.** 600   **5.** 0.1862   **6.** 400

7. 2222   **8.** 0.4065   **9.** 4800   **10.** 10.8   **11.** 0.0062

12. 90,000   **13.** 0.387   **14.** 730.9   **15.** 83.9   **16.** 7840.6

17. 89.2644   **18.** 780   **19.** 31

Exercises

Fill in the blanks.

1. 5,470 centigrams = _____ grams

2. 14,016 meters = _____ kilometers

3. 481 l = _____ hl

4. 0.036 hectoliter = _____ deciliters

5. 36 g = _____ hg

6. 1861 decimeters = _____ hectometers

7. 7.11 dal = _____ dl

8. 727.7 milligrams = _____ decigrams

9. 873 deciliters = _____ hectoliter

10. 0.9043 hg = _____ cg

11. 47 deciliters = _____ milliliters

12. 1.793 dam = _____ cm

13. 99.07 dg = _____ cg

14. 0.1 kiloliter = _____ milliliters

15. 0.00841 kilogram = _____ decigrams

16. 0.0845 meter = _____ millimeters

17. 18 mm = _____ cm

18. 6.7 dal = _____ cl

19. 0.8 cl = _____ ml

20. 36,771 mg = _____ dag

Answers

1. 54.7	2. 14,016	3. 4.81	4. 36	5. 0.36
6. 1.861	7. 711	8. 7.277	9. 0.873	10. 9.043
11. 4,700	12. 1,793	13. 990.7	14. 100,000	15. 84.1
16. 84.5	17. 1.8	18. 6,700	19. 8	20. 3.6771

Quick Metric Facts

You don't have to spend hours trying to convert

> miles to kilometers
>
> pounds to grams
>
> gallons to liters, or
>
> Fahrenheit to Celsius temperatures.

Here are a few quick facts to make the job easier.

Kilometers to Miles

Suppose you are driving in Canada, where speed limits are given in kilometers per hour. Your speedometer measures miles per hour. You look in the rearview mirror and see a police car.

Don't panic.

Here's a quick way to convert kilometers per hour into miles per hour.

Example   The Canadian speed limit sign says 80 kilometers per hour. Multiply the first digit, 8, by 6.

$$8 \times 6 = 48$$

You need to drive no faster than 48 miles per hour to avoid speeding.

Another
Example

The speed limit sign says 50 kilometers per hour. What is the speed limit in miles per hour? Multiply the first digit, 5, by 6.

$$5 \times 6 = 30$$

The speed limit is about 30 mph.

Grams to Pounds

Suppose you are still in Canada, and you stop at a supermarket to buy some groceries. All the packages are labeled in metric units.

Example

A package of bacon weighs 500 g (grams). How much does it weigh in the English system?

Think of 500 g as a nice, fat pound. A pound is actually 454 g, but usually you won't need to be that accurate.

Another
Example

A package of sliced meat weighs 125 g. How much does it weigh in pounds?

Think of this amount as roughly 125 g out of 500 g (which is about a pound).

$$\frac{\overset{1}{\cancel{125}}}{\underset{4}{\cancel{500}}} = \frac{1}{4} \text{ of a pound}$$

Instead of dividing everything out, make a "guesstimate," and you'll get pretty close to the answer.

Liters to Quarts

A quart is about the same size as a liter. Unless you're buying something really expensive, think of 1 liter as equal to 1 quart.

$$3 \text{ liters of milk} = 3 \text{ quarts of milk}$$

Gasoline Mileage

Metric mileage is measured in liters of gas per 100 kilometers of driving. Trying to convert liters per kilometer into miles per gallon gets involved. Let's see if we can speed up the process.

Example   You have driven 200 miles since you filled your gas tank, and it takes 40 liters of gas to fill the tank again. What mileage are you getting?

Step 1   Change liters to quarts:

40 liters = 40 quarts (approximately).

Step 2   Change quarts to gallons:

Since there are 4 quarts in each gallon,

$$40 \text{ quarts} = \frac{\overset{10}{\cancel{40}}}{\underset{1}{\cancel{4}}} \text{ or 10 gallons.}$$

Step 3   Divide miles by gallons:

$$\text{Your car mileage is } \frac{\overset{20}{\cancel{200} \text{ miles}}}{\underset{1}{\cancel{10} \text{ gallons}}}, \text{ or 20 miles per gallon.}$$

You are getting 20 miles per gallon of gas.

Another Example   You have driven 300 miles on 50 liters of gas. What is the mileage?

Step 1   50 liters = 50 quarts

Step 2   50 quarts = $\frac{50}{4}$ or   12.5 gallons

Step 3   Your mileage is $\frac{300 \text{ miles}}{12.5 \text{ gallons}}$ or 24 miles per gallon.

Temperature Conversions

Temperatures in the United States are measured in degrees Fahrenheit, and temperatures in Canada are measured in degrees Celsius. It takes too long to convert temperatures from one measuring system to the other.

An easy way to figure out temperatures is to think of room temperature, which is about 22°C (Celsius).

Examples
If a Toronto weather station says the temperature is 25°C, you know we are having a heat wave.

If the weather forecaster says the temperature is 18°C, take your sweater with you.

If you look out your window in the morning and see a heavy dew on your car, wear a light coat. The temperature is around 5°C to 10°C and quite chilly.

If the dew on your car has turned to frost, you guessed it. The temperature is 0°C or lower and freezing.

That frost on your car might make you wish for those pleasant summer days when you were outside in your shirt sleeves. The temperature then ranged from 20°C to about 24°C.

If the forecaster says it's going to rain, take your umbrella. Metric rain hasn't changed from the old system. It's still wet!

If the weather report says that 30 millimeters of rain fell, don't run for the lifeboats. 30 millimeters is only about 1 inch.

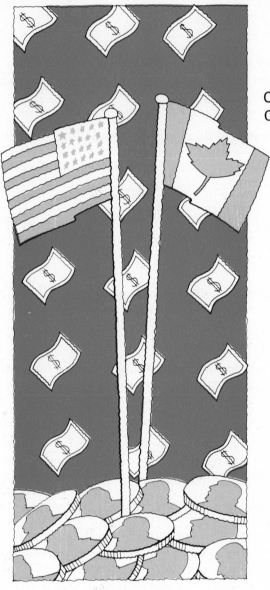

One thing that the United States and Canada have in common is MONEY.

Both systems are metric. Take a look at this:

10 pennies	=	1 dime
10 dimes	=	1 dollar
10 ones	=	1 ten-dollar bill
10 tens	=	1 hundred-dollar bill
10 hundreds	=	1 thousand
1000 thousand	=	1 million (If you've got it, I'm in love!)

We've been metric all along, but we never thought of it that way.

Try the review problems on the next page.

Review Exercises

Fill in the blanks for Exercises 1–3.

| 60 kph | = | ? mph | | 90 kph | = | ? mph | | 100 kph | = | ? mph |

1. _____

2. _____

3. _____

4. On a winter day the thermometer in your living room shows a temperature of 24°C. If you want to adjust the temperature to an average room temperature, would you turn the thermostat up or down?

5. A jar of peanut butter contains 900 grams. Is this about 1 pound, 2 pounds, or 3 pounds?

6. You just bought a 4-liter can of maple syrup. How many gallons of syrup did you buy?

7. 80 millimeters of snow fell last night. Very roughly, how many inches is this?

8. You are used to the metric system, because you've been using it for years. In what form?

9. The service station attendant fills your tank, and the pump shows 60 liters. About how many gallons is this?

10. A wedge of cheese is marked 500 grams. How many pounds does the cheese weigh?

Unit 12 Test

Use a separate piece of paper to work out each problem.

Give the meaning of each of the following prefixes.
(For example, *deka* means 10.)

1. deci   **2.** centi   **3.** hecto   **4.** milli

Put the metric abbreviations in the correct places on this line.

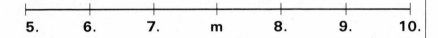

5.   **6.**   **7.**   m   **8.**   **9.**   **10.**

Fill in the missing numbers.

11. 5 m = _____ mm   **12.** 5 m = _____ km

13. 1.8 dm = _____ dam   **14.** 0.2 hm = _____ cm

Put the metric abbreviations in the correct places on the
following two lines. Fill in the blanks for problems 21 and
22.

15.   **16.**   **17.**   l   **18.**   **19.**   **20.**

21. 24 kl = _____ ml   **22.** 32 cl = _____ hl

23.   **24.**   **25.**   g   **26.**   **27.**   **28.**

Fill in the blanks.

29. 64.5 dag = _____ g   **30.** 8 g = _____ kg

31. 8 g = _____ dg   **32.** 1 mg = _____ kg

33. 250 mg = _____ lb   **34.** 70 kph = _____ mph

35. You are in Toronto to see the White Sox play the Blue
Jays. A sign in the stadium parking lot gives a speed
limit of 20 kilometers per hour. How many miles per
hour is this?

Your Answers

1. _____
2. _____
3. _____
4. _____
5. _____ 6. _____
7. _____ 8. _____
9. _____ 10. _____
11. _____
12. _____
13. _____
14. _____
15. _____ 16. _____
17. _____ 18. _____
19. _____ 20. _____
21. _____
22. _____
23. _____ 24. _____
25. _____ 26. _____
27. _____ 28. _____
29. _____
30. _____
31. _____
32. _____
33. _____
34. _____
35. _____

13 | Geometry of Straight Lines

Aims you toward:

- Knowing a little about lines and angles.

- Knowing six different geometric shapes.

- Knowing how to find perimeter.

- Knowing how to find area.

- Knowing how to find volume.

In this unit you will work exercises like these:

- Give the name and perimeter of this figure.

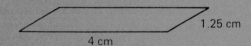

1.25 cm

4 cm

- Give the area of this figure.

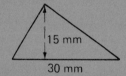

15 mm

30 mm

- Give the volume of this figure.

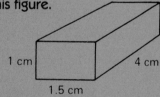

1 cm

4 cm

1.5 cm

Lines and Angles

All circles are divided into 360 degrees. This is usually written 360°.

Sailors imagine 360 circles drawn through the north and south poles, and use these circles to tell where they are.

Half a circle is

$$\frac{1}{2} \times 360°.$$

$$\frac{1}{\overset{2}{\cancel{2}}}_{1} \times \frac{\overset{180°}{\cancel{360°}}}{1} = 180°$$

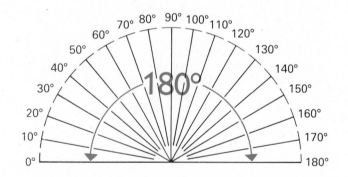

Since a straight line divides the circle in half, it is also 180°.

Sometimes a straight line is also called a STRAIGHT ANGLE, which sounds a little strange.

When it comes right down to it, you can't really see a line, because a line is actually the DISTANCE BETWEEN TWO POINTS.

In other words, a line has length, but no width. You can't really draw a line, because the width of your pencil mark makes it a long, thin bar rather than a true line.

All this is just theory, because everyone uses the expression "draw a line," and that's pretty hard to do without a pencil!

Back to degrees:

If you divide a straight angle of 180° in half, you have

$$180° \div 2 = 90°.$$

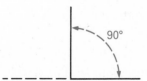

A 90° angle is also called a RIGHT ANGLE, and is shown with a little box in the corner.

Whenever you see an angle with a little box in the corner, you know it is a 90° angle, or right angle.

Four-sided figures with corners of 90° are called SQUARES (if all sides are the same length) or RECTANGLES (if there are 2 long sides and 2 short sides).

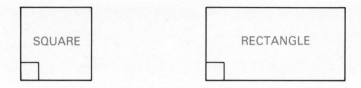

Triangles

A tricycle has 3 wheels, and *tri* means 3. A TRIANGLE has 3 angles.

If you have a triangle where one of the angles is a right angle, you have a RIGHT TRIANGLE.

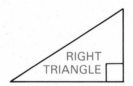

In a little while you'll be working with lines that are at right angles to each other, so you'll need to know two things:

The line that goes up is called HEIGHT.

The line that is flat is called the BASE.

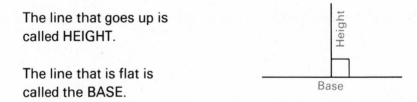

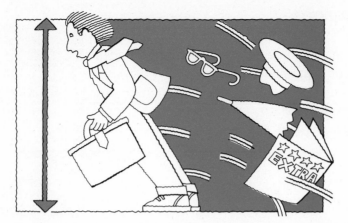

This man walking against the wind may be 6 feet tall, but his real height now is only 5 feet because the top of his head is only 5 feet from the ground, or BASE line.

Similarly, the sides of this figure might be 6 feet long, but its real height is only 5 feet.

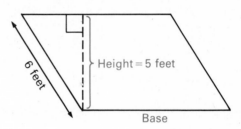

Exercises

1. How many degrees are in a circle?

2. How many degrees in a straight line?

3. A line is the _____ between _____ _____ . (Fill in the blanks.)

4. How many degrees in a right angle?

5. What is the symbol for a right angle? (Draw it if you wish.)

6. What is the name of this figure?

7. Name line A and line B in the figure at right.

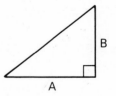

Answers

1. 360° 2. 180° 3. Distance, two points 4. 90° 5. A little box
6. Right triangle 7. Line A is the base, and line B is the height.

Geometric Shapes

This section focuses on some different geometric shapes. But first, look at some parallel lines.

Like railway tracks, parallel lines run the same way and never cross.

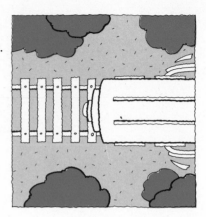

You can show that two lines are parallel with arrows.

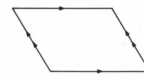

Two *more* sides are parallel.

Two sides are parallel.

Now look at the following examples of geometric shapes.

QUADRILATERAL

Any 4-sided figure is a quadrilateral.

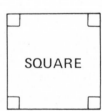

SQUARE

A square always has 4 right angles and 4 equal sides.

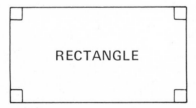

RECTANGLE

A rectangle has 4 right angles, and its opposite sides are equal.

PARALLELOGRAM

The opposite sides of a parallelogram are equal, and its opposite sides are parallel.

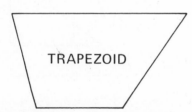

TRAPEZOID

Two opposite sides of a trapezoid are parallel, while the other sides are not parallel.

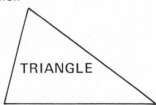

TRIANGLE

Any 3-sided figure is a triangle.

Perimeter

If you were asked to find the number of feet of sidewalk needed to go around a city block, you would just add up all the sides.

700 feet + 750 feet + 700 feet + 750 feet = 2,900 feet

You would need 2,900 linear feet of sidewalk.

The distance around the outside of any geometric figure is called the PERIMETER of the figure. To find the perimeter, add up the lengths of all the sides.

Example   Find the perimeter of this figure.

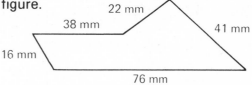

38 mm + 22 mm + 41 mm + 76 mm + 16 mm = 193 mm

Another Example   What is the perimeter of this rectangle?

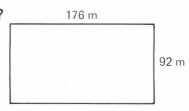

You know that two of the sides are 176 meters and 92 meters. Since this is a rectangle, you know that the other two sides are also 176 meters and 92 meters.

176 m + 92 m + 176 m + 92 m = 536 m

Exercises

Name each of the following shapes and find their perimeters.

1.
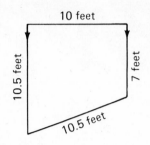
10 feet
10.5 feet
7 feet
10.5 feet

2.
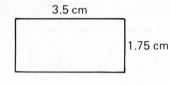
3.5 cm
1.75 cm

3.
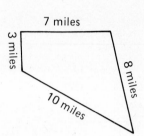
7 miles
3 miles
8 miles
10 miles

4.
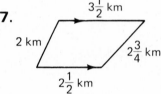
13 yards

5.

$2\frac{1}{2}$ cm
3 cm
$1\frac{1}{2}$ cm

6.
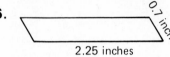
0.7 inches
2.25 inches

7.

$3\frac{1}{2}$ km
2 km
$2\frac{3}{4}$ km
$2\frac{1}{2}$ km

8.
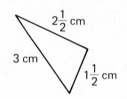
30 mm

9.
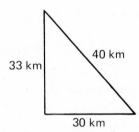
33 km
40 km
30 km

10.
40 mm
20 mm

11.
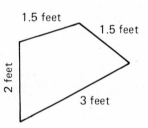
1.5 feet
1.5 feet
2 feet
3 feet

12.
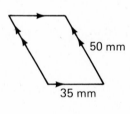
50 mm
35 mm

Answers

10. Rectangle, 120 mm	**11.** Quadrilateral, 8 feet	**12.** Parallelogram, 170 mm
7. Trapezoid, $10\frac{3}{4}$ km	**8.** Square, 120 mm	**9.** Triangle, 103 km
4. Square, 52 yards	**5.** Triangle, 7 cm	**6.** Parallelogram, 5.9 inches
1. Trapezoid, 38 feet	**2.** Rectangle, 10.5 cm	**3.** Quadrilateral, 28 miles

Area

AREA is a measure of the whole *surface* of any figure. Look at this rectangle.

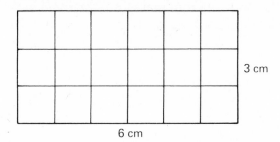

3 cm

6 cm

Two sides are 6 centimeters long. The other sides are 3 centimeters long. Inside the rectangle there are 18 squares, each with sides of 1 centimeter. This rectangle has an area of 18 square centimeters (18 cm²).

1 cm

1 cm

Area is measured in terms of square centimeters (cm²), square inches (inches²), square kilometers (km²), square feet (feet²), and so on. When you say, for instance, that a field is 10 square kilometers (10 km²) you mean that 10 squares, each with sides of 1 kilometer, would fit exactly into the field.

Are you wondering about the raised 2 in things like km²? A 2 used in that way means that whatever you have is multiplied by itself. For example, $8^2 = 8 \times 8 = 64$. As another example, $3^2 = 3 \times 3 = 9$. We read 3^2 as "three squared," or "three to the second power." km² means km $\times$ km, and is read "kilometers squared." You will see below why you need the raised 2 for area.

An easy way to find the area of many figures is to multiply the BASE times the HEIGHT.

AREA = BASE $\times$ HEIGHT

The BASE is one of the sides. For RECTANGLES and SQUARES, the HEIGHT is just another side.

Example   Find the area of this rectangle.

2 feet } Height

4 feet

Base

Area = Base $\times$ Height
= 4 feet $\times$ 2 feet
= 8 feet²

Here you are multiplying $4 \times 2 = 8$. You are also multiplying feet $\times$ feet = feet². So the answer is 8 feet². (Remember, *always* use a raised 2 for AREA.)

To find the area of a PARALLELOGRAM, you use the same formula:
AREA = BASE × HEIGHT. But the HEIGHT is not merely one of the sides.

Height is always measured on a line that makes a *right angle* with the base. The dotted
lines below show the height of each parallelogram.

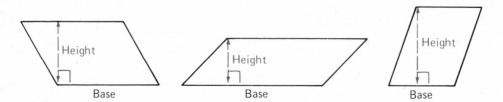

Why is height measured like this? The area of a parallelogram is the same as the area
of the rectangle you would make from it, if you cut off one corner at a right angle and
moved it over to the other end.

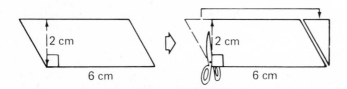

The area of this rectangle,
and therefore the area of the
parallelogram, is 6 cm × 2 cm.

So when you figure the area of a parallelogram (other than a rectangle), never use the
length of a side for height. A side does not make a right angle with the base.

Example   What is the area of this parallelogram?

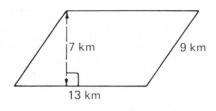

Area = Base × Height
     = 13 km × 7 km
     = 91 km²

Remember: the height is 7 km, not 9 km.

Exercises on finding area are on the next page.

Exercises

Find the area for each of the following figures.

1.

30 mm

2.

30 mm

50 mm

3.

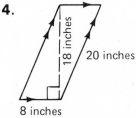

4 cm

$4\frac{1}{2}$ cm

5 cm

4.

18 inches

20 inches

8 inches

5.

11 feet

6.

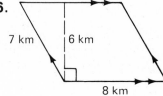

7 km   6 km

8 km

Give the *name,* the *perimeter,* and the *area* for each of the following figures.

7.

10 mm

20 mm

8.

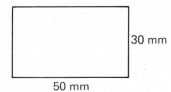

12 mm

9.

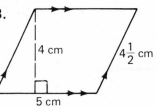

13 mm   10 mm

20 mm

10.

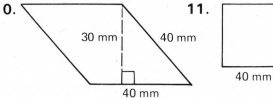

30 mm   40 mm

40 mm

11.

40 mm

40 mm

12.

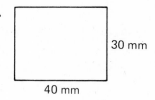

30 mm

40 mm

Answers

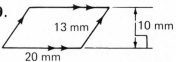

1. 900 mm²   2. 1500 mm²   3. 20 cm²
4. 144 inches²   5. 121 feet²   6. 48 km²
7. Rectangle, 60 mm, 200 mm²   8. Square, 48 mm, 144 mm²
9. Parallelogram, 66 mm, 200 mm²   10. Parallelogram, 160 mm, 1200 mm²
11. Square, 160 mm, 1600 mm²   12. Rectangle, 140 mm, 1200 mm²

Area of Triangles

Look again at the area of a rectangle.

Area = Base × Height
= 6 cm × 4 cm
= 24 cm²

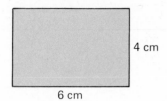

4 cm

6 cm

If this rectangle is cut in half as shown, it becomes two triangles.

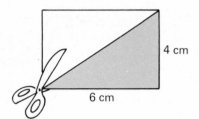

4 cm

6 cm

So to find the area of each triangle, you take half of the area of the rectangle.

$$\text{AREA OF A TRIANGLE} = \frac{\text{BASE} \times \text{HEIGHT}}{2}$$

Example Find the area of these two figures.

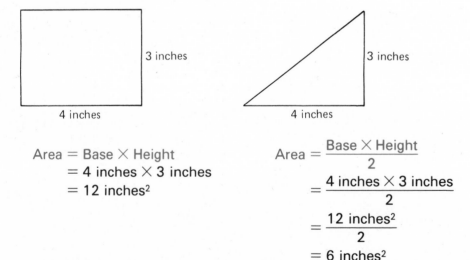

3 inches

4 inches

3 inches

4 inches

Area = Base × Height
= 4 inches × 3 inches
= 12 inches²

$$\text{Area} = \frac{\text{Base} \times \text{Height}}{2}$$
$$= \frac{4 \text{ inches} \times 3 \text{ inches}}{2}$$
$$= \frac{12 \text{ inches}^2}{2}$$
$$= 6 \text{ inches}^2$$

The triangle has half the area of the rectangle.

As you learned earlier in this unit, triangles like the ones on this page are called RIGHT TRIANGLES. Right triangles have one angle that is a right angle. One of the sides attached to the right angle is the *base.* The other side attached to the right angle is the *height.*

What about triangles that do not have a right angle?

Look at the area of the parallelogram.

Area = Base × Height
$$= 34 \text{ mm} \times 20 \text{ mm}$$
$$= 680 \text{ mm}^2$$

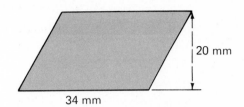

20 mm

34 mm

If this parallelogram is cut in half like this, it becomes two triangles.

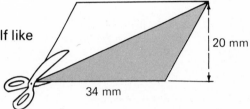

20 mm

34 mm

So to find the area of each triangle you take half of the area of the parallelogram.

$$\textbf{AREA OF A TRIANGLE} = \frac{\textbf{BASE} \times \textbf{HEIGHT}}{\textbf{2}}$$

Example   Find the area of these two figures.

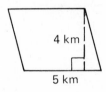

4 km

5 km

4 km

5 km

Area = Base × Height
$$= 5 \text{ km} \times 4 \text{ km}$$
$$= 20 \text{ km}^2$$

$$\text{Area} = \frac{\text{Base} \times \text{Height}}{2}$$
$$= \frac{5 \text{ km} \times 4 \text{ km}}{2}$$
$$= \frac{20 \text{ km}^2}{2}$$
$$= 10 \text{ km}^2$$

The triangle has half the area of the parallelogram.

All triangles use the same formula for area: $\textbf{AREA OF A TRIANGLE} = \dfrac{\textbf{BASE} \times \textbf{HEIGHT}}{\textbf{2}}$.

Right triangles use a side for the height.

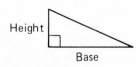

Height

Base

If a triangle does not have a right angle, the height is found by a line that makes a right angle with the base and connects the base to the top point of the triangle.

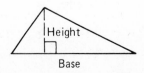

Height

Base

Exercises

Find the *area* for each of the following figures.

1.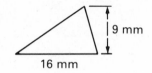
8 mm
20 mm

2.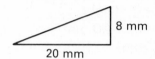
8 mm
20 mm

3.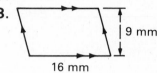
9 mm
16 mm

4.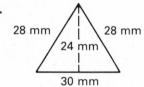
9 mm
16 mm

5.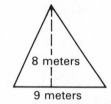
8 meters
9 meters

6.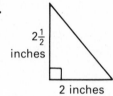
$2\frac{1}{2}$ inches
2 inches

7.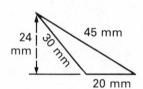
28 mm   28 mm
24 mm
30 mm

8.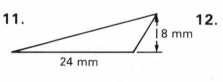
24 mm   45 mm   30 mm
20 mm

9.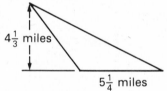
$4\frac{1}{3}$ miles
$5\frac{1}{4}$ miles

10.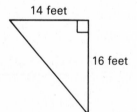
14 feet
16 feet

11.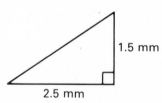
8 mm
24 mm

12.
1.5 mm
2.5 mm

Answers

1. 160 mm²	**2.** 80 mm²	**3.** 144 mm²	**4.** 72 mm²
5. 36 meters²	**6.** $2\frac{1}{2}$ inches²	**7.** 360 mm²	**8.** 240 mm²
9. $11\frac{3}{8}$ miles²	**10.** 112 feet²	**11.** 96 mm²	**12.** 1.875 mm²

Area of Trapezoids

To find the area of a trapezoid (a quadrilateral with only two opposite sides parallel) you need to know the base and the height just as you do in the other types of figures.

For trapezoids, *the height is the distance between the two parallel sides.* If one of the sides forms right angles with the two parallel lines, then that side is the height. Otherwise, you need a line that forms a right angle with the two parallel lines to find the height. The dotted lines in the figures below show the height for each trapezoid.

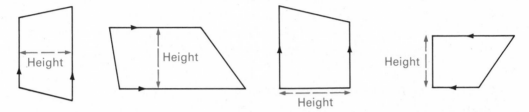

As you have learned, the height forms a right angle with the base. With trapezoids, the height forms a right angle with 2 different lines.

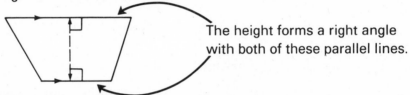

The height forms a right angle with both of these parallel lines.

You can't use both of these lines for the base, so you AVERAGE the lengths of the two parallel lines.

Use this formula to find the area of a trapezoid:

AREA OF A TRAPEZOID = AVERAGE BASE × HEIGHT.

Example   Find the area of this trapezoid.

8 cm
7 cm
10 cm

Height = 7 cm

To find the Average Base add 8 cm + 10 cm, then divide by 2.

$$\text{Average Base} = \frac{8 \text{ cm} + 10 \text{ cm}}{2} = \frac{18 \text{ cm}}{2} = 9 \text{ cm}$$

Area = Average Base × Height
     = 9 cm × 7 cm
     = 63 cm²

Don't worry, there are more examples before the exercises!

Example   What is the area of this trapezoid?

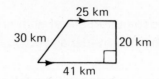

Height = 20 km

Average Base = $\dfrac{41\ km + 25\ km}{2} = \dfrac{66\ km}{2} = 33\ km$

Area = Average Base × Height
33 km × 20 km
660 km²

Another
Example   Find the area of this trapezoid.

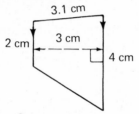

Height = 3 cm

Average Base = $\dfrac{4\ cm + 2\ cm}{2} = \dfrac{6\ cm}{2} = 3\ cm$

Area = Average Base × Height
= 3 cm × 3 cm
= 9 cm²

Another
Example   Find the area of this trapezoid.

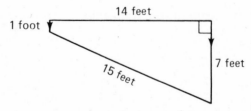

Height = 14 feet (It may help to turn the page sideways to see this.)

Average Base = $\dfrac{1\ foot + 7\ feet}{2} = \dfrac{8\ feet}{2} = 4\ feet$

Area = Average Base × Height
= 4 feet × 14 feet
= 56 feet²

Now you're ready to work the exercises.

Exercises

Find the *area* of each of these trapezoids.

1.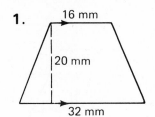
16 mm
20 mm
32 mm

2.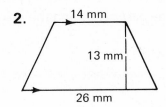
14 mm
13 mm
26 mm

3.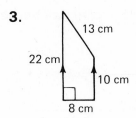
13 cm
22 cm
10 cm
8 cm

4.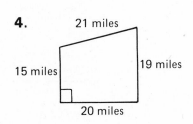
21 miles
15 miles
19 miles
20 miles

5.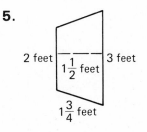
2 feet
3 feet
$1\frac{1}{2}$ feet
$1\frac{3}{4}$ feet

6.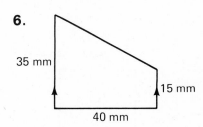
35 mm
15 mm
40 mm

7.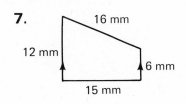
16 mm
12 mm
6 mm
15 mm

8.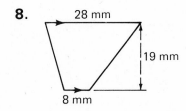
28 mm
19 mm
8 mm

9.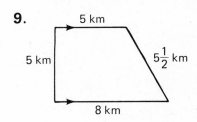
5 km
5 km
$5\frac{1}{2}$ km
8 km

10.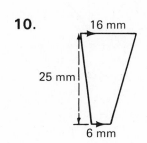
16 mm
25 mm
6 mm

11.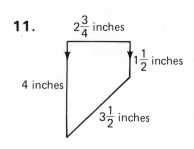
$2\frac{3}{4}$ inches
$1\frac{1}{2}$ inches
4 inches
$3\frac{1}{2}$ inches

12.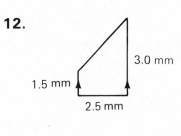
3.0 mm
1.5 mm
2.5 mm

Answers

1. 480 mm² **2.** 260 mm² **3.** 128 cm² **4.** 340 miles²

5. $3\frac{3}{4}$ feet² **6.** 1000 mm² **7.** 135 mm² **8.** 342 mm²

9. $32\frac{1}{2}$ km² **10.** 275 mm² **11.** $7\frac{9}{16}$ inches² **12.** 5.625 mm²

These two pages cover all the geometric shapes you have learned about so far.

Review Exercises

For each figure give the *name,* the *perimeter,* and the *area.*

1.
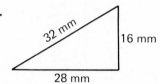

2.
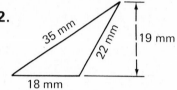

3.
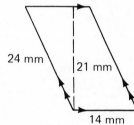

4.
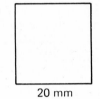

5.
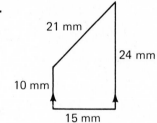

6.
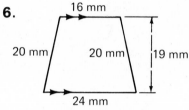

7.
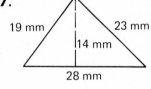

8.

9.
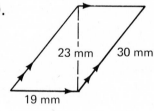

Answers

9. Parallelogram, 98 mm, 437 mm²
7. Triangle, 70 mm, 196 mm²
5. Trapezoid, 70 mm, 255 mm²
3. Parallelogram, 76 mm, 294 mm²
1. Triangle, 76 mm, 224 mm²

8. Rectangle, 84 mm, 320 mm²
6. Trapezoid, 80 mm, 380 mm²
4. Square, 80 mm, 400 mm²
2. Triangle, 75 mm, 171 mm²

10.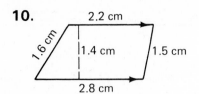
2.2 cm
1.6 cm
1.4 cm
1.5 cm
2.8 cm

11.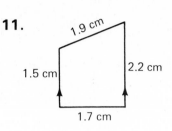
1.9 cm
1.5 cm
2.2 cm
1.7 cm

12.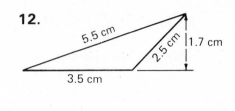
5.5 cm
2.5 cm
1.7 cm
3.5 cm

13.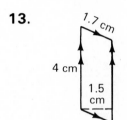
1.7 cm
4 cm
1.5 cm

14.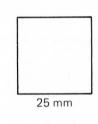
1.6 cm

15.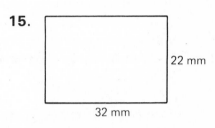
22 mm
32 mm

16.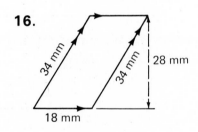
34 mm
34 mm
28 mm
18 mm

17.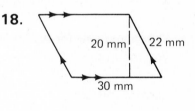
25 mm

18.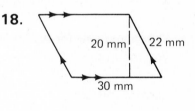
20 mm
22 mm
30 mm

19.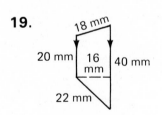
18 mm
20 mm
16 mm
40 mm
22 mm

20.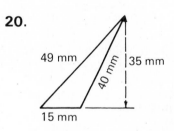
49 mm
40 mm
35 mm
15 mm

21.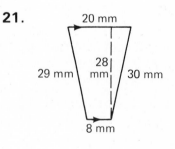
20 mm
29 mm
28 mm
30 mm
8 mm

Volume

Area is a measure of things that are flat.

VOLUME is the measurement of things that are three-dimensional.

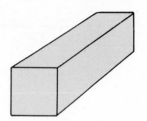

Volume is measured in terms of CUBES. A CUBE is a solid box that has the same measurements on each side. Here is a CUBIC CENTIMETER.

Each edge of a cubic centimeter is 1 centimeter long.

Volume can be measured in cubic centimeters (cm³), cubic millimeters (mm³), cubic inches (inches³), cubic feet (feet³), and so on.

When you say, for example, that an object is 12 cubic millimeters (12 mm³) in volume, you mean that 12 cubes with edges of 1 millimeter can fit exactly into the object.

Look at this box.

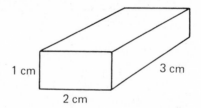

It is 2 centimeters wide, 1 centimeter high, and 3 centimeters long.

Here the box has been divided into cubes.
There are 6 cubes each with edges of 1 centimeter.

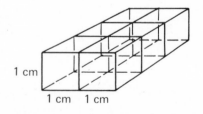

The box is 6 cubic centimeters (6 cm³) in volume.

That's not magic; that's arithmetic!

> The raised 3 in things like mm³ means that whatever you have is multiplied by itself, then multiplied by itself again. So 2³, "two cubed," or "two to the third power," means 2 × 2 × 2 or 8, and 3³ means 3 × 3 × 3 or 27. Also, mm³ means mm × mm × mm.

The next page shows how to find volume using formulas.

You can use two steps to find the volumes of some kinds of objects. First you find the area of one of the ends, then you multiply the area by the length of the object.

Example Find the volume of this object.

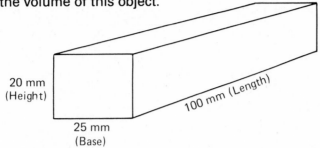

20 mm
(Height)

25 mm
(Base)

100 mm (Length)

Formula: Volume = Base × Height × Length

Step 1: Base × Height = 25 mm × 20 mm

= 500 mm² (Area of end)

Step 2: Volume = 500 mm² × 100 mm (Length)

= 50,000 mm³

Why is the answer given in mm³? Because you are multiplying mm × mm × mm, which is mm³.

Example What is the volume of this object?

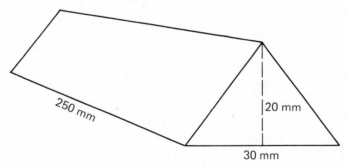

250 mm

20 mm

30 mm

Formula: Volume = $\dfrac{\text{Base} \times \text{Height}}{2}$ × Length

Step 1: $\dfrac{\text{Base} \times \text{Height}}{2} = \dfrac{30 \text{ mm} \times 20 \text{ mm}}{2}$

$= \dfrac{600 \text{ mm}^2}{2}$

= 300 mm² (Area of end)

Step 2: Volume = 300 mm² × 250 mm (Length)

= 75,000 mm³

Exercises

Find the *volume* of each of the following objects.

Hint: First find the area of the end, then multiply by the length.

1.

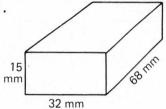

15 mm
68 mm
32 mm

2.

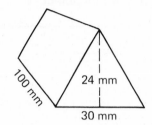

100 mm
24 mm
30 mm

3.

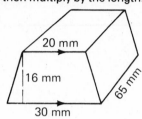

20 mm
16 mm
65 mm
30 mm

4.

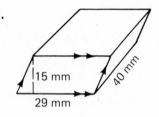

15 mm
40 mm
29 mm

5.

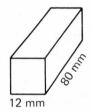

80 mm
12 mm

6.

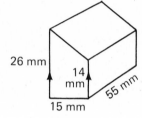

26 mm
14 mm
55 mm
15 mm

7.

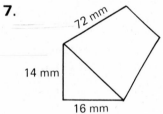

72 mm
14 mm
16 mm

8.

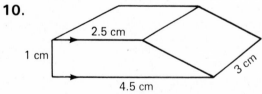

2 inches
4 inches
10 inches

9.

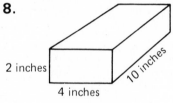

3 feet
3¾ feet
2 feet
5 feet

10.

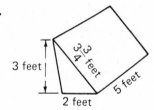

2.5 cm
1 cm
3 cm
4.5 cm

11.

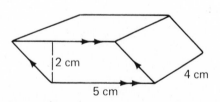

2 cm
5 cm
4 cm

These exercises cover all the material in this unit. Last chance to practice before the test!

Review Exercises

Give the *perimeter* and the *area* of the shapes in Problems 1–4.

1.

18 mm 23 mm
32 mm

2.

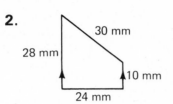

30 mm
28 mm
10 mm
24 mm

3.

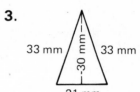

33 mm 33 mm
30
21 mm

4.

22 mm 17 mm
12 mm
36 mm

5. How many cubic inches of water will a rectangular fish tank hold, if the width is 18 inches, the height is 14 inches, and the length is 24 inches?

Find the *volume* for each of the following objects.

6.

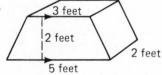

3 feet
2 feet
2 feet
5 feet

7.

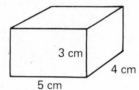

3 cm
4 cm
5 cm

8.

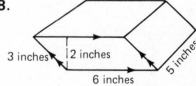

3 inches 2 inches
6 inches
5 inches

9.

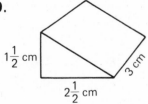

$1\frac{1}{2}$ cm
3 cm
$2\frac{1}{2}$ cm

10. How many cubic feet of cement are in a cubic yard? (*Hint:* Start with the number of feet in a yard.)

Answers

9. $5\frac{5}{8}$ cm³   **10.** 27 feet³

6. 16 feet³   **7.** 60 cm³   **8.** 60 inches³

4. 75 mm, 216 mm²   **5.** 6048 inches³

1. 110 mm, 576 mm²   **2.** 92 mm, 456 mm²   **3.** 87 mm, 315 mm²

Unit 13 Test

Use a separate piece of paper to work out each problem.

Give the *name* and *perimeter* of each of the following shapes.

1.

4 meters

2.5 meters

2.

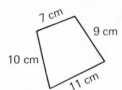

7 cm

9 cm

10 cm

11 cm

3.

2 miles

1 mile   1.75 miles

2.75 miles

4.

9 km

Give the *area* of each of the following shapes.

5.

80 mm

100 mm

6.

36 cm

63 cm

7.

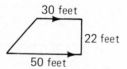

30 feet

22 feet

50 feet

8.

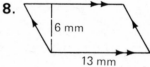

6 mm

13 mm

9.

11 inches

20 inches

10.

2 m

4 m   2.5 m

Give the *volume* of each of the following objects.

11.

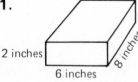

2 inches

6 inches

8 inches

12.

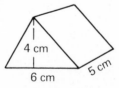

4 cm

6 cm

5 cm

13.

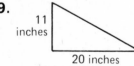

2 inches

3.5 inches

7 inches

14.

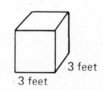

3 feet

3 feet

15.

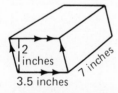

19 mm

8 mm

16 mm

15 mm

16.

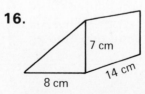

7 cm

8 cm

14 cm

Your Answers

1. _____
2. _____
3. _____
4. _____
5. _____
6. _____
7. _____
8. _____
9. _____
10. _____
11. _____
12. _____
13. _____
14. _____
15. _____
16. _____

14 | Geometry of Circles

Aims you toward:

- Knowing the meaning of radius, diameter, and circumference.

- Knowing how to find circumference.

- Knowing how to find the area of a circle.

- Knowing how to find area and volume for combined shapes.

In this unit you will work exercises like these:

- What is the circumference of this circle?

- What is the area of this circle?

- What is the volume of this cylinder?

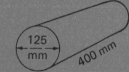

Circles

CIRCLES have been making life easier for a long time.

If you're going to talk about circles, there are a few words you'll need to know.

The RADIUS is the distance halfway across a circle, or from the edge to the center.

The DIAMETER is the distance all the way across a circle through the center.

The CIRCUMFERENCE is the distance around the outside of a circle. The circumference is the same as the perimeter.

Thousands of years ago mathematicians learned that if the circumference of a circle is about 22 units (meters, inches, or whatever), then the diameter will be about 7 units.

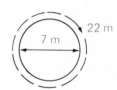

Since the ratio of the circumference to the diameter is the same for all circles, there is a special name for it. The ratio was named π (pronounced *pie*) for a letter in the Greek alphabet.

π is equal to about $\frac{22}{7}$.

In decimal form π is equal to about 3.14.

$$
\begin{array}{r}
3.14 \\
7\overline{)22.00} \\
21 \\
\hline
1\,0 \\
7 \\
\hline
30 \\
28 \\
\hline
2
\end{array}
$$

π will keep coming up when you work with circles, so remember that π is equal to about $\frac{22}{7}$ or about 3.14.

Some exercises are next.

Exercises

1. What fraction is an approximation for π?

2. What decimal is an approximation for π?

Give the name of the distance shown by the dotted line in each circle below.

3.   **4.**   **5.**

Give the *radius* of each of the following circles.

6.   **7.**   **8.**

7 cm   18 inches   50 mm

Give the *diameter* of each of the following circles.

9.   **10.**   **11.**

2 m $2\frac{1}{2}$ km $1\frac{3}{4}$ feet

Give the *radius* of each of the following circles

12.   **13.**   **14.**

1.5 yd $1\frac{1}{2}$ cm   2.04 inches

Answers

1. $\frac{22}{7}$	**2.** 3.14	**3.** Circumference	**4.** Diameter
5. Radius	**6.** 3.5 cm	**7.** 9 inches	**8.** 25 mm
9. 4 m	**10.** 5 km	**11.** $3\frac{1}{2}$ feet	**12.** 0.75 yards
13. 0.75 cm	**14.** 1.02 inches		

Circumference

You know that a circle with a diameter of about 7 units has a circumference of about 22 units. To find the circumferences of other circles, multiply the diameter by π.

CIRCUMFERENCE = π × DIAMETER

You can use either $\frac{22}{7}$ or 3.14 for π. Both give an approximate answer that is close enough for the problems in this book. The answers given for the exercises in this book use $\pi = 3.14$.

Example   What is the circumference of this circle?

$$\text{Circumference} = \pi \times \text{Diameter}$$
$$= 3.14 \times 21 \text{ inches}$$
$$= 65.94 \text{ inches}$$

Another Example   Find the distance around this figure (half circle).

First think of a whole circle and find the circumference.

$$\text{Circumference} = \pi \times \text{Diameter}$$
$$= \frac{22}{7} \times 14 \text{ cm}$$
$$= \frac{22}{7} \times \frac{14}{1} \text{ cm}$$
$$= 44 \text{ cm}$$

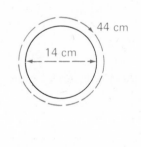

Now take half of this circumference.

$$44 \div 2 = 22 \text{ cm}$$

You now have this much of the figure.

Add on the diameter.

$$22 \text{ cm} + 14 \text{ cm} = 36 \text{ cm}$$

The distance around the figure is 36 cm.

Now work the exercises on circumference.

Exercises

Find the circumference of each of the following circles. (Remember use $\pi = 3.14$ if the problem has decimals in it, and $\frac{22}{7}$ if you're working with fractions.)

1.
4 feet

2.
24 inches

3.
2 m

Careful! 2 m is the radius. You need the diameter.

4.
35 mm

5.
56 mm

6.
42 mm

7.
1.8 cm

8.
$\frac{3}{4}$ m

9.
$1\frac{1}{4}$ mm

Find the distance around each of the following figures.

10.
10 feet

11.
11 cm

12.
5m

Answers

12. 7.85 m (half circle) + 5 m (diameter) = 12.85 m
11. 17.27 cm (half circle) + 11 cm (diameter) = 28.27 cm
10. 15.7 feet (half circle) + 10 feet (diameter) = 25.7 feet

9. $3\frac{13}{14}$ mm   **8.** $4\frac{5}{7}$ m   **7.** 11.304 cm

6. 131.88 mm   **5.** 175.84 mm   **4.** 109.9 mm

3. 12.56 m   **2.** 75.36 inches   **1.** 12.56 feet

Exercises

Find the circumference of each of the following figures. (Use $\pi = 3.14$.) The first one is done to get you started.

1.

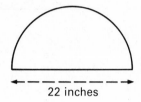

22 inches

If it was a whole circle, $C = \pi d$ or 3.14×22. But, C of only half a circle is $\dfrac{3.14 \times 22}{2}$ or 34.54 inches. Don't forget to add the length of the bottom line: $34.54 + 22 = 56.54$ inches.

2.

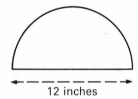

12 inches

3.

5 yards

4.

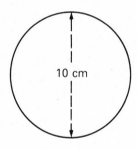

10 cm

5.

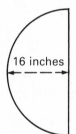

16 inches

6. The trunk of a walnut tree measures 38 inches in diameter. What is its circumference?

7. Find the approximate circumference of the rim of a bicycle wheel if the length of a spoke is 12 inches.

8. A circular flower bed has a diameter of 14 feet. How many blocks, each 1 foot long, would be needed to go around the bed? (Round off to the nearest whole number.)

Answers

1. 56.54 inches   2. $18.84 + 12 = 30.84$ inches   3. 15.7 yards
4. 31.4 cm   5. $50.24 + 32 = 82.24$ inches   6. 119.32 inches
7. 75.36 inches   8. 44 blocks

Area of a Circle

The AREA of a circle is the whole surface.

To find the AREA of a circle, use this formula.

$$\textbf{AREA OF A CIRCLE} = \pi \times \textbf{RADIUS}^2$$

Remember that a raised 2 means something is multiplied by itself. So the above formula is actually

$$\textbf{AREA OF A CIRCLE} = \pi \times \textbf{RADIUS} \times \textbf{RADIUS.}$$

"Radius $\times$ Radius" is written "Radius^2" because it is shorter.

The area of a circle is measured in square units just as any other area.

Example   What is the area of this circle?

$$
\begin{aligned}
\text{Area} &= \pi \times \text{Radius}^2 \\
&= \pi \times \text{Radius} \times \text{Radius} \\
&= \pi \times 21 \text{ mm} \times 21 \text{ mm} \\
&= \pi \times 441 \text{ mm}^2 \\
&= 3.14 \times 441 \text{ mm}^2 \\
&= 1384.74 \text{ mm}^2
\end{aligned}
$$

Another   Find the area of this circle.
Example

$$\text{Area} = \pi \times \text{Radius}^2$$

To find the radius, divide the diameter by 2.

$$56 \div 2 = 28$$

$$
\begin{aligned}
\text{Area} &= \pi \times \text{Radius} \times \text{Radius} \\
&= \pi \times 28 \text{ mm} \times 28 \text{ mm} \\
&= \pi \times 784 \text{ mm}^2 \\
&= 3.14 \times 784 \text{ mm}^2 \\
&= 2461.76 \text{ mm}^2
\end{aligned}
$$

More exercises are on the next page.

Exercises

Find the area for each of the following. (Use $\pi = 3.14$.) Round off to the nearest hundredth.

1.

2.

8.4 mm

3.

63 mm

You want radius, not diameter!

4.

35 mm

5.

7 inches

6.

12.6 mm

Just find the area of a circle with radius 12.6 and divide the area in half.

7.

14 mm

8.

21 mm

You want radius, not diameter!

9.

40.6 mm

10.

19 feet

11.

30.8 mm

12.

16.8 mm

Answers

1. 153.86 mm²	**2.** 221.56 mm²	**3.** 3115.67 mm²
4. 3846.50 mm²	**5.** 153.86 inches²	**6.** 249.25 mm²
7. 615.44 mm²	**8.** 173.09 mm²	**9.** 1293.96 mm²
10. 141.69 feet²	**11.** 372.34 mm²	**12.** 110.78 mm²

Volume

To find out how much water a pipe like this will hold you need to know its VOLUME.

$$\text{VOLUME} = \pi \times \text{RADIUS}^2 \times \text{LENGTH}$$

The volume of an object with circular ends is measured in cubic units just as any other volume, and you find it in the same way. First you find the area of one of the ends, and then you multiply the area by the length of the object.

Example   What is the volume of this cylinder?

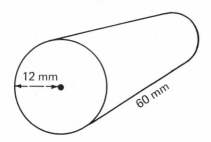

Formula: Volume $= \pi \times \text{Radius}^2 \times \text{Length}$

Step 1: $\pi \times \text{Radius}^2 = \pi \times 12\text{ mm} \times 12\text{ mm}$

$= \pi \times 144\text{ mm}^2$

$= 3.14 \times 144\text{ mm}^2$

$= 452.16\text{ mm}^2$ (Area of end)

Step 2:   Volume $= 452.16\text{ mm}^2 \times 60\text{ mm}$ (Length)

$= 27,129.60\text{ mm}^3$

Exercises on volume are next.

Exercises

Find the volume of each of the following objects. (Use $\pi = 3.14$.) Round off to the nearest hundredth.

1.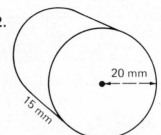
13 cm
26 cm

2.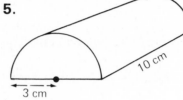
20 mm
15 mm

3.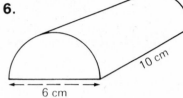
8 m
14 m

4.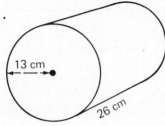
2 inches
5 inches

5.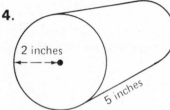
10 cm
3 cm

6.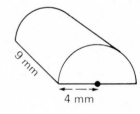
10 cm
6 cm

7.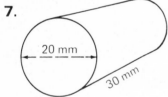
20 mm
30 mm

8.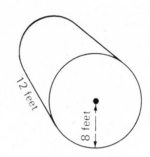
12 feet
8 feet

9.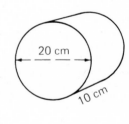
9 mm
4 mm

10.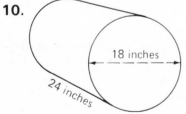
18 inches
24 inches

11.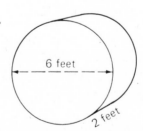
6 feet
2 feet

12.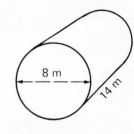
20 cm
10 cm

Answers

1. 13797.16 cm³ **2.** 18840 mm³ **3.** 703.36 m³ **4.** 62.8 inches³
5. 141.3 cm³ **6.** 141.3 cm³ **7.** 9420 mm³ **8.** 2411.52 feet³
9. 226.08 mm³ **10.** 6104.16 inches³ **11.** 56.52 feet³ **12.** 3140 cm³

These exercises cover all the material in this unit. Here are the circle formulas you will need.

$$\textbf{CIRCUMFERENCE} = \pi \times \textbf{DIAMETER}$$

$$\textbf{AREA} = \pi \times \textbf{RADIUS}^2$$

$$\textbf{VOLUME} = \pi \times \textbf{RADIUS}^2 \times \textbf{LENGTH}$$

Be sure to MEMORIZE these formulas if you haven't already. You need to know them for the test.

Review Exercises

1. What fraction is an approximation for π?

2. What decimal is an approximation for π?

Give the name of the distances shown by the dotted lines in these circles.

3. 　　　**4.** 　　　**5.**

Give the *diameter* of each circle.

6.
4 cm

7.
20 mm

8.
22 inches

Find the *circumference* of each of the following circles. Round off to the nearest hundredth.

9.
14 cm

10.
20 feet

11.
5 mm

Answers

	11. 31.40 mm	10. 62.80 feet	9. 43.96 cm
8. 7 inches	7. 40 mm	6. 8 cm	5. Radius
4. Diameter	3. Circumference	2. 3.14	1. $\frac{22}{7}$

Give the *distance around* each of the following figures. (Use $\pi = 3.14$.) Round off to the nearest hundredth.

12.
10 inches

13.
14 inches

14.
6 inches

Give the *area* of each of the following figures.

15.
8 mm

16.
2 cm

17.
24 yards

Give the *volume* of each of the following figures.

18.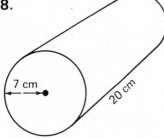
7 cm
20 cm

19.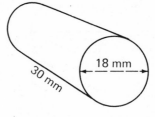
30 mm
18 mm

20.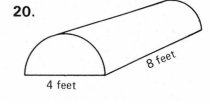
4 feet
8 feet

21.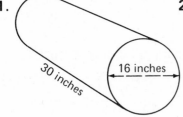
30 inches
16 inches

22. A dairy butter churn is a revolving cylinder, about 12 feet long and 5 feet in diameter. In order to thrash the cream properly, the cylinder is always operated $\frac{1}{3}$ full. This allows the cream to be churned into butter. How many cubic feet of cream are used for each batch of butter?

Answers

12. 31.4 inches	**13.** 35.98 inches	**14.** 30.84 inches
15. 200.96 mm²	**16.** 3.14 cm³	**17.** 904.32 yards
18. 3077.2 cm³	**19.** 7630.20 mm³	**20.** 50.24 feet³
21. 6028.80 inches³	**22.** 78.5 feet³	

Unit 14 Test

Use a separate piece of paper to work out each problem.

Give the *circumference* and *area* of each of the following. (Use $\pi = 3.14$.) Round off to the nearest hundredth.

1.

26 cm

2.

12 mm

3.

5 inches

4.

6 mm

Give the *distance around* and the *area* for each of the following. Round off to the nearest hundredth.

5.

11 cm

6.

73 mm

7. John ordered a large pizza, 16 inches in diameter. The pizza parlor sent over 2 smaller pizzas, each 8 inches in diameter. They told John that he was getting the same amount of pizza, since the two smaller diameters equaled the larger ($8 + 8 = 16$). John complained, saying he wasn't getting as much. Was he right?

Give the *volume* for each of the following. Round off to the nearest hundredth.

8.
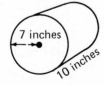
7 inches
10 inches

9.
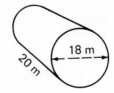
18 m
20 m

10.
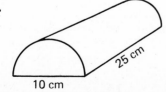
25 cm
10 cm

Your Answers

1. _____

_____

2. _____

_____

3. _____

_____

4. _____

_____

5. _____

_____

6. _____

_____

7. _____

_____

8. _____

_____

9. _____

_____

10. _____

_____

Review Test

This test covers all the geometry you have learned.

Give the name, the *perimeter* (or *circumference*), and the *area* for each of the following. (Use $\pi = 3.14$.) Round off to the nearest hundredth if needed.

1.
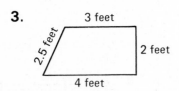
12 inches
3 inches
4 inches

2.

5.5 cm
3 cm

3.
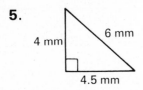
3 feet
2.5 feet
2 feet
4 feet

4.

8 m

5.
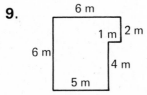
6 mm
4 mm
4.5 mm

6.
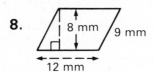
15 cm
14 cm
15 cm
14 cm

Give the *perimeter* and the *area* for each of the following. Round off to the nearest hundredth if needed.

7.
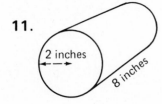
9 m

8.
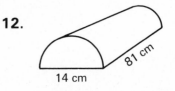
8 mm
9 mm
12 mm

9.
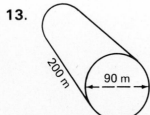
6 m
1 m
2 m
6 m
4 m
5 m

10.
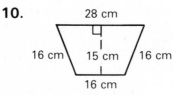
28 cm
16 cm
15 cm
16 cm
16 cm

Give the *volume* for each of the following.

11.
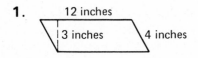
2 inches
8 inches

12.
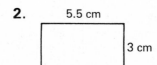
81 cm
14 cm

13.
200 m
90 m

14. What is the volume of a can that is 2 inches in diameter and 4 inches high?

Your Answers

1. _____

   _____

2. _____

   _____

3. _____

   _____

4. _____

   _____

5. _____

   _____

6. _____

   _____

7. _____

   _____

8. _____

   _____

9. _____

   _____

10. _____

   _____

11. _____

   _____

12. _____

   _____

13. _____

   _____

14. _____

   _____

15 | Combined Operations

Aims you toward:

- Knowing the order of operations.

- Knowing how to find the average for a group of numbers.

In this unit you will work exercises like these:

- Solve: $43 - 23 \times 4 \div 3 + 4$.

- Find the average for these numbers: 6 4 7 5.

Combined Operations

You now know how to perform the four operations—addition, subtraction, multiplication, and division—with three different types of numbers: whole numbers, fractions, and decimals. That's quite an accomplishment!

Here's a problem that puts some of these skills together.

$$9 - 5 + 4 \times 8$$

You work the problem like this:

$9 - 5 + 4 \times 8$ Multiply first: $4 \times 8 = 32$,

$9 - 5 + 32$ then subtract: $9 - 5 = 4$,

$4 + 32$ and finally add: $4 + 32 = 36$.

36

Mix up the signs and you work the problem this way:

$9 + 5 \times 4 - 8$ Multiply first: $5 \times 4 = 20$,

$9 + 20 - 8$ then add: $9 + 20 = 29$,

$29 - 8$ and finally subtract: $29 - 8 = 21$.

21

Mix up the signs again and you work it this way:

$9 \times 5 - 4 + 8$ Multiply first: $9 \times 5 = 45$,

$45 - 4 + 8$ then subtract: $45 - 4 = 41$,

$41 + 8$ and finally add: $41 + 8 = 49$.

49

How did we know when to add, when to subtract, and when to multiply? Got the feeling that everybody else is in on some secret that you don't know about?

Combined Operations: Addition and Subtraction

Here's the secret. . . .

There are rules everybody follows that tell the order for working a problem involving more than one operation.

Rule 1　When working a problem involving ADDITION and/or SUBTRACTION, do the operations in the order they are written, from left to right.

Example　$6 - 2 + 3$

$6 - 2 + 3$　　Subtraction comes first, so do it first: $6 - 2 = 4$.

$4 + 3$　　Then add: $4 + 3 = 7$.

7　　The answer is 7.

Another Example　$6 + 2 - 3$

$6 + 2 - 3$

$6 + 2 - 3$　　Addition comes first, so do it first.

$8 - 3$

5

Another Example　$6 + 2 + 3 + 4 + 1$

$6 + 2 + 3 + 4 + 1$

$8 \quad + 3 + 4 + 1$

$11 \quad + 4 + 1$

$15 \quad + 1$

16

Now try some exercises.

Exercises

Work each of the following problems. Keep in mind the first rule about order of operations.

1. $10 + 2 - 3 - 1 + 4 - 2$

2. $15 - 2 + 6 + 14$

3. $6 + 7 + 4 + 10$

4. $3 - 2 - 1 + 5 - 3 - 2 + 10$

5. $3 + 6 - 4 - 2 + 8 - 1$

6. $6 - 5 + 7 - 3$

7. $98 - 50 + 101$

8. $6 - 0 + 4 - 1 + 0 - 2 + 3$

9. $14 - 1 + 3 + 2 - 3 + 1 - 6$

10. $46 + 8 - 13 - 5$

11. $18 - 11 + 11$

12. $1 + 4 - 2 - 1 + 7 - 1 + 2$

The same order of operations rule is used in problems involving addition and subtraction of decimals or fractions.

Example $0.014 + 6.3 - 4.102 + 8.2 - 0.412$

$\underbrace{0.014 + 6.3} - 4.102 + 8.2 - 0.412$

$\underbrace{6.314 - 4.102} + 8.2 - 0.412$

$\underbrace{2.212 + 8.2} - 0.412$

$\underbrace{10.412 - 0.412}$

10.000

Remember that with fractions you need to change the fractions to equivalent fractions with common denominators before you can add or subtract.

Example $\dfrac{3}{4} - \dfrac{1}{2} + \dfrac{1}{8} - \dfrac{1}{16}$

$\underbrace{\dfrac{3}{4} - \dfrac{1}{2}} + \dfrac{1}{8} - \dfrac{1}{16}$ First make equivalent fractions for $\dfrac{3}{4}$ and $\dfrac{1}{2}$.

$\underbrace{\dfrac{3}{4} - \dfrac{2}{4}} + \dfrac{1}{8} - \dfrac{1}{16}$ Then subtract.

$\underbrace{\dfrac{1}{4} + \dfrac{1}{8}} - \dfrac{1}{16}$ Now you need equivalent fractions for $\dfrac{1}{4}$ and $\dfrac{1}{8}$.

$\underbrace{\dfrac{2}{8} + \dfrac{1}{8}} - \dfrac{1}{16}$ Add.

$\underbrace{\dfrac{3}{8} - \dfrac{1}{16}}$ Make equivalent fractions for $\dfrac{3}{8}$ and $\dfrac{1}{16}$.

$\underbrace{\dfrac{6}{16} - \dfrac{1}{16}}$ Subtract.

$\dfrac{5}{16}$ The answer is $\dfrac{5}{16}$.

Example $1\dfrac{1}{3} + 2\dfrac{1}{2} - \dfrac{1}{4}$

$\underbrace{1\dfrac{1}{3} + 2\dfrac{1}{2}} - \dfrac{1}{4}$ First make equivalent fractions for $\dfrac{1}{3}$ and $\dfrac{1}{2}$.

$\underbrace{1\dfrac{2}{6} + 2\dfrac{3}{6}} - \dfrac{1}{4}$ Add.

$\underbrace{3\dfrac{5}{6} - \dfrac{1}{4}}$ Make equivalent fractions for $\dfrac{5}{6}$ and $\dfrac{1}{4}$.

$\underbrace{3\dfrac{10}{12} - \dfrac{3}{12}}$ Subtract.

$3\dfrac{7}{12}$ The answer is $3\dfrac{7}{12}$.

Exercises

Remember the first rule about order of operations when you work these problems.

1. $1 - \dfrac{1}{6} - \dfrac{1}{4} + \dfrac{1}{3}$

2. $3.4 - 1.02 + 10.18 - 3.4$

3. $3.008 + 11 - 0.004 - 9.1$

4. $72 - 43.06 + 0.24 - 19.18$

5. $2\dfrac{1}{6} + \dfrac{1}{3} + \dfrac{1}{4} - 1\dfrac{1}{2} - \dfrac{4}{5}$

6. $2 - 1\dfrac{7}{8} + \dfrac{3}{4} - \dfrac{1}{16}$

7. $72.4 + 0.18 - 31.9$

8. $7\dfrac{1}{2} - \dfrac{3}{5} + \dfrac{1}{4} - 3\dfrac{2}{5}$

9. $42.8 + 0.109 - 6$

10. $\dfrac{1}{6} + \dfrac{4}{5} + \dfrac{1}{2} - \dfrac{2}{3}$

11. $11\dfrac{5}{8} - 2\dfrac{3}{4} - 6\dfrac{1}{2}$

12. $2.111 + 4.082 - 1.106 + 12.3 - 7.387$

Answers

1. $\dfrac{11}{12}$ 2. 9.16   3. 4.904   4. 10.0   5. $\dfrac{9}{20}$

6. $\dfrac{13}{16}$ 7. 40.68   8. $\dfrac{3}{4}$ 9. 36.909   10. $\dfrac{4}{5}$

11. $2\dfrac{3}{8}$ 12. 10.0

Combined Operations: Multiplication and Division

The rule for working problems involving MULTIPLICATION and/or DIVISION is much the same as the rule for addition and subtraction.

Rule 2   When a problem has multiplication and/or division, do them in the order they appear, from left to right.

Example $10 \div 2 \times 4 \div 10$

$10 \div 2 \times 4 \div 10$ Division comes first, so do it first.

$5 \times 4 \div 10$ Then multiply.

$20 \div 10$ Divide again.

2 The answer is 2.

Another Example $45 \times 2 \div 3 \times 4$

$45 \times 2 \div 3 \times 4$ This time multiplication comes first.

$90 \div 3 \times 4$

30×4

120

Another Example $36 \times 4 \times 2 \times 3$

$36 \times 4 \times 2 \times 3$

$144 \times 2 \times 3$

288×3

864

MULTIPLY OR DIVIDE FROM LEFT TO RIGHT

Now try the exercises. ▶

Exercises

1. $8 \times 3 \div 6 \div 2 \times 3$

2. $6 \times 3 \times 2 \div 12 \times 2 \div 3 \times 4$

3. $15 \div 5 \times 2 \div 3 \times 6 \div 3$

4. $42 \div 3 \times 6 \times 2 \div 4$

5. $12 \div 3 \times 2 \div 4$

6. $99 \div 11 \times 7 \div 3 \div 7$

7. $84 \div 4 \times 5 \times 2 \div 7$

8. $8 \times 3 \div 4 \times 2 \div 6$

9. $69 \div 3 \times 4 \div 2$

10. $14 \div 2 \times 4 \times 3 \div 42$

Answers

1. 6   2. 8   3. 4   4. 42   5. 2   6. 3   7. 30   8. 2   9. 46   10. 2

Here are some more problems involving multiplication and division, but with decimals and fractions instead of whole numbers.

Example $17.6 \div 4 \times 36.2 \div 0.2$

$17.6 \div 4 \times 36.2 \div 0.2$ Watch the decimal places.

$\underbrace{4.4 \times 36.2} \div 0.2$

$\underbrace{159.28 \div 0.2}$

796.4

Fractions are a little more difficult. You must be careful to invert the proper fractions when you divide.

Example $\dfrac{1}{2} \times \dfrac{1}{4} \div \dfrac{3}{8} \times 2$

$\underbrace{\dfrac{1}{2} \times \dfrac{1}{4}} \div \dfrac{3}{8} \times 2$ First multiply.

$\underbrace{\dfrac{1}{8} \div \dfrac{3}{8}} \times 2$ To divide, invert $\dfrac{3}{8}$ and multiply.

$\underbrace{\dfrac{1}{\overset{}{\underset{1}{\cancel{8}}}} \div \dfrac{\overset{1}{\cancel{8}}}{3}} \times 2$

$\underbrace{\dfrac{1}{3} \times 2}$ Multiply again.

$\underbrace{\dfrac{1}{3} \times \dfrac{2}{1}}$

$\dfrac{2}{3}$ The answer is $\dfrac{2}{3}$.

Another Example $\dfrac{2}{3} \div \dfrac{1}{4} \times \dfrac{1}{2}$

$\underbrace{\dfrac{2}{3} \div \dfrac{1}{4}} \times \dfrac{1}{2}$

$\underbrace{\dfrac{2}{3} \times \dfrac{4}{1} \times \dfrac{1}{2}}$

$\underbrace{\dfrac{\overset{4}{\cancel{8}}}{3} \times \dfrac{1}{\underset{1}{\cancel{2}}}}$

$\dfrac{4}{3}$ or $1\dfrac{1}{3}$

Exercises

1. $18.3 \div 0.3 \times 0.2 \div 4$

2. $2\frac{1}{4} \times \frac{1}{8} \div 1\frac{3}{4} \times 12\frac{4}{9} \times 3$

3. $56 \div 5 \times 0.04 \div 0.7$

4. $1\frac{1}{5} \div \frac{1}{2} \times 1\frac{1}{4} \div 3 \times 5$

5. $400.1 \div 50 \times 30 \div 6$

6. $14 \times \frac{1}{5} \div \frac{7}{15} \div 3 \times \frac{1}{2}$

7. $1\frac{3}{10} \times 2\frac{2}{3} \div \frac{3}{5} \times 1\frac{1}{2} \times \frac{3}{13}$

8. $15 \div 6 \times 0.3 \div 0.4$

9. $4 \div \frac{1}{2} \times 2\frac{1}{2} \div 2 \times \frac{1}{5} \times 4$

10. $186 \div 0.4 \div 30$

Answers

1. 3.05   2. 6   3. 0.64   4. 5   5. 40.01
6. 1   7. 2   8. 1.875   9. 8   10. 15.5

Combined Operations: Order of Operations

Rule 1 and Rule 2 work for some problems, but not all of them. What about problems that have BOTH addition or subtraction AND multiplication or division? We need a special rule for these, or else we could end up with two different answers to the same problem!

$$5 + 2 \times 2$$

$$7 \times 2$$

$$14$$

If we add first, the answer is 14.

$$5 + 2 \times 2$$

$$5 + 4$$

$$9$$

If we multiply first, the answer is 9.

We can't have TWO answers for ONE problem!

Most countries of the world have agreed to work problems the same way. That way, all of us can get the same answers.

All of this brings us to Rule 3.

Rule 3   When addition and/or subtraction along with multiplication and/or division are all in the same problem, do the multiplication and division first from left to right. Then do the addition and subtraction from left to right.

Example $3 + 4 \div 2 - 2 \times 2$

$$3 + 4 \div 2 - 2 \times 2$$
$$3 + 2 - 2 \times 2$$
$$3 + 2 - 4$$
$$5 - 4$$
$$1$$

Start at the left and look for multiplication or division. Division comes first, so do it first.
Continue looking for multiplication or division. Multiplication is next.
There is no more multiplication or division, so begin back at the left and do the addition and subtraction.
The answer is 1.

Another Example $4.08 \times 2 - 3 \div 0.5 \times 1.1$

$$4.08 \times 2 - 3 \div 0.5 \times 1.1$$
$$8.16 - 3 \div 0.5 \times 1.1$$
$$8.16 - 6 \times 1.1$$
$$8.16 - 6.6$$
$$1.56$$

Another Example $\frac{1}{4} + \frac{1}{2} \times \frac{1}{4} - \frac{1}{8} \div 2$

$\frac{1}{4} + \underbrace{\frac{1}{2} \times \frac{1}{4}} - \frac{1}{8} \div 2$ First multiply.

$\frac{1}{4} + \frac{1}{8} - \underbrace{\frac{1}{8} \div 2}$ Then divide.

$\frac{1}{4} + \frac{1}{8} - \underbrace{\frac{1}{8} \times \frac{1}{2}}$

$\underbrace{\frac{1}{4} + \frac{1}{8}} - \frac{1}{16}$ Then add.

$\underbrace{\frac{2}{8} + \frac{1}{8}} - \frac{1}{16}$

$\underbrace{\frac{3}{8} - \frac{1}{16}}$ Finally, subtract.

$\underbrace{\frac{6}{16} - \frac{1}{16}}$

$\frac{5}{16}$

In summary, here are the three order of operation rules.

Order of Operations	1. If a problem involves only addition and/or subtraction, work the problem in order from left to right.
	2. If a problem involves only multiplication and/or division, work the problem in order from left to right.
	3. If a problem involves addition and/or subtraction as well as multiplication and/or division, do the multiplication and division first from left to right. Then do the addition and subtraction from left to right.

Exercises

The exercises on this page cover many of the topics that have been discussed so far in this text.

1. $8 - 6 \div 3 + 3 \times 2 - 11$

2. $49 \times 0.1 + 18 \div 2 - 6$

3. $86 - 7 \times 4.1 + 15 \div 5$

4. $\frac{3}{8} \times 4.14 \div \frac{1}{2} - 0.43$

5. $\frac{1}{3} + \frac{1}{2} \div \frac{1}{4} - \frac{1}{8} \times 2 - 1\frac{1}{12}$

6. $8 - 2 \times 3 + 4 \div 4$

7. $3.68 \div 0.4 - 0.113 + 6$

8. $4 + 3 \times 2 - 8 \div 2 - 5$

9. $\frac{1}{8} \div 2 \div \frac{7}{16} - \frac{1}{9}$

10. $8 - \frac{1}{4} \times 3 \div \frac{1}{2} - 5\frac{1}{2}$

Answers

1. 1 **2.** 7.9 **3.** 60.3 **4.** 2.675 **5.** 1 **6.** 3 **7.** 15.087 **8.** 1 **9.** $\frac{2}{63}$ **10.** 1

Averaging

A market researcher conducted a telephone survey to study how many weekly magazines people buy. She recorded the number of different magazines in each household at the time of her call. She ended up with the following list.

5 1 0 3 2 4 1 2 0 2 4 1 3 5 7 4 2 1 0 3 4

By itself this list doesn't mean very much. To make it more meaningful, the market researcher found its AVERAGE.

In order to find the average for a group of numbers, you use a combination of addition, counting, and division.

Example   Find the average for these numbers: 4   7   6   8   5.

$$\begin{array}{r} 4 \\ 7 \\ 6 \\ 8 \\ +\ 5 \\ \hline 30 \end{array}$$

First add all the numbers together.

Next count how many numbers there are to be averaged. Here there are 5 numbers.

$$\begin{array}{r} 6 \\ 5\overline{)30} \\ \underline{30} \\ 0 \end{array}$$

Divide 5 into 30.
The average for the numbers is 6.

Another Example   Find the average for these numbers: $\frac{1}{2}$ $\frac{5}{8}$ $\frac{3}{4}$ $\frac{1}{4}$ $\frac{3}{8}$.

$$\frac{4}{8} + \frac{5}{8} + \frac{6}{8} + \frac{2}{8} + \frac{3}{8} = \frac{20}{8}$$

First make equivalent fractions and add.

$$\frac{20}{8} \div 5$$

There are 5 numbers, so divide by 5.

$$\frac{\overset{4}{\cancel{20}}}{8} \times \frac{1}{\underset{1}{\cancel{5}}} = \frac{4}{8} = \frac{1}{2}$$

The average of the numbers is $\frac{1}{2}$.

To Find the Average	Add the numbers. Count how many numbers there are to be averaged. Divide the sum by the amount of numbers being averaged.

If you are finding the average of whole numbers or decimals, write the answer as a whole number or a decimal. If you are finding the average of fractions or mixed numbers, write the answer as a fraction or a mixed number.

Example   During a 4-year period from 1979 to 1982, winning horses in the Triple Crown races brought in the following prize money: $1,279,334 won by Spectacular Bid, $1,130,450 won by Temperance Hill, $1,148,800 won by John Henry, and $1,197,400 won by Penault. What was the average amount of prize money won by these horses?

$$\begin{array}{r}
\$1{,}279{,}334 \\
\$1{,}130{,}450 \\
\$1{,}148{,}800 \\
+\ \$1{,}197{,}400 \\
\hline
\$4{,}755{,}984
\end{array}$$ First add up all the numbers.

$$\begin{array}{r}
1{,}188{,}996 \\
4\overline{)4{,}755{,}984} \\
\underline{4} \\
7 \\
\underline{4} \\
35 \\
\underline{32} \\
35 \\
\underline{32} \\
39 \\
\underline{36} \\
38 \\
\underline{36} \\
24 \\
\underline{24} \\
0
\end{array}$$ Since there are 4 numbers, we divide by 4.

The average amount won is $1,188,996.

Try this problem on finding average.

The number of people per square mile in each province of Canada is listed below. Find the average number of people per square mile in the whole country. (*Hint:* Don't forget —when no decimal is showing, it is understood to be at the end of the number. A number such as 42 is the same as 42.0.) British Columbia: 7.8 persons per square mile, Manitoba: 4.9 persons per square mile, New Brunswick: 25.4, Newfoundland: 4, Nova Scotia: 42, Ontario: 25.4, Prince Edward Island: 56.7, Quebec: 12.4, Saskatchewan: 4.5, Alberta: 9.4, Yukon: 0.1, and the Northwest Territories: 0.06.

Answers

192.66 ÷ 12 = 16.055

Exercises

Find the average of each of the following lists of numbers. Round decimals to the nearest hundredth if needed.

1. 16 14 10 11 13

2. 150 148

3. 76.4 75.7 74.3 75.6

4. 3 $3\frac{1}{2}$ $2\frac{5}{8}$ $3\frac{1}{4}$

5. 8 7.89 7.43 8.1

6. 9.9 10.3 9.8

7. $\frac{7}{13}$ $\frac{9}{26}$ $\frac{1}{2}$

8. 21 22 24 20 23

9. What is your test average if you receive the following test scores in one semester?
75 82 85 83 79 86

10. The daily receipts for one week at a fast-food restaurant were $1,768.12, $1,412.46, $1,089.73, $1,586.63, $1,986.40, $1,821.43, and $1,470.38. What was the average daily cash intake for the week?

11. Suppose the price of round trip bus fare between Chicago and Kansas City increased $2.00 in 1979, $2.10 in 1980, $1.30 in 1981, $2.80 in 1982, and $2.15 in 1983. What would be the average yearly increase for these years?

12. The earth's atmosphere has a layer of gas around it called ozone. This is a type of oxygen, and is 12 miles above the earth in some places. In other areas it is 21 miles up, but can be as low as 6 miles and as high as 35 miles. What is the average height of the ozone layer?

Answers

1. 12.8   **2.** 149   **3.** 75.5   **4.** $3\frac{3}{32}$ **5.** 7.86

6. 10   **7.** $\frac{9}{13}$ **8.** 22   **9.** 81.67

10. $1,590.74   **11.** $2.07   **12.** 18.5 miles

Unit 15 Test

Use a separate piece of paper to work out each problem.

1. $6 - 2 - 3 - 1 + 2$

2. $\frac{1}{6} + 1\frac{1}{2} - \frac{1}{3} - 1\frac{1}{4}$

3. $1\frac{2}{3} \div 5 \times 1\frac{1}{6}$

4. $0.52 \div 0.04 \times 0.012$

5. $\frac{1}{2} + \frac{1}{4} \times \frac{1}{3} - \frac{1}{8} \div \frac{1}{4}$

6. $42 - 6 + 5 \times 2 + 0.14 \times 8$

7. $12 - 2.04 + 3 \times 4 - 13.72$

Find the average of each of the following lists of numbers. Round off decimals to the nearest thousandth if needed.

8. $\frac{1}{8}$ $\frac{1}{4}$ $\frac{3}{16}$ $\frac{5}{16}$ $\frac{3}{8}$

9. 36   34   35   36

10. $\frac{5}{9}$ $\frac{1}{2}$ $\frac{2}{3}$ $\frac{1}{2}$

11. 2.746   2.103   2.696

Work each of these word problems.

12. A swimmer beginning to train for the Olympics records her time for the 100-meter freestyle. She gets the following times: 95.6 seconds, 75.4 seconds, 86.3 seconds, and 88.2 seconds. What was her average time for these trials? Round off to the nearest tenth.

13. Each of three different automotive service stations charge the following for an oil change and tune-up: $60.40, $48.50, $73.80. What is the average charge?

14. Oakgrove Cinema's grand opening of a new movie lasted four days. They sold 439 tickets the first day, 532 the second day, 486 the third day, and 510 the fourth day. What was the average for these days? Round off to the nearest whole number.

Your Answers

1. _____

2. _____

3. _____

4. _____

5. _____

6. _____

7. _____

8. _____

9. _____

10. _____

11. _____

12. _____

13. _____

14. _____

16 | Preparing for Algebra: Order of Operations

Aims you toward:

- Reviewing the basic rules for order of operations.
- Knowing how to use exponents.
- Knowing how to use parentheses in arithmetic expressions.
- Knowing how to solve involved fractions.

In this unit you will work exercises like these:

- Solve: $4^2 + 3(5 - 3)^3$.

- Solve: $\dfrac{3 + 4(2) - 1}{2(3) + 2 - 3}$.

Order of Operations

Don't be scared by the word ALGEBRA! The arithmetic you have been studying is one branch of mathematics. Algebra is just another branch.

In algebra you need the rules about order of operations that you learned in Unit 15. Here they are again.

Order of Operations	1. If a problem involves only addition and/or subtraction, work the problem in order from left to right.
	2. If a problem involves only multiplication and/or division, work the problem in order from left to right.
	3. If a problem involves addition and/or subtraction as well as multiplication and/or division, do the multiplication and division from left to right. Then do the addition and subtraction from left to right.

Example $3 + 7 \times 8 - 5$

$3 + 7 \times 8 - 5$

$3 + 56 - 5$

$59 - 5$

54

Another Example $16 \div 8 \times 3 + 7 - 3 \times 2$

$16 \div 8 \times 3 + 7 - 3 \times 2$

$2 \times 3 + 7 - 3 \times 2$

$6 + 7 - 3 \times 2$

$6 + 7 - 6$

$13 - 6$

7

Exercises

Keep in mind the rules about order of operations as you work each of the following.

1. $14 \div 7 + 1 - 4 \times 0.5$

2. $9 \times 2 \div 3 \div 2 + 3$

3. $47 - 17 + 6 - 4$

4. $8.4 - 5.6 + 16$

5. $\dfrac{1}{2} + \dfrac{1}{4} - \dfrac{3}{4} + 7$

6. $\dfrac{2}{3} \times \dfrac{1}{5} + \dfrac{3}{30} - \dfrac{2}{15}$

7. $42 \div 7 + 8 \times 3 - 19$

8. $22 \div 11 \times 14 + 3 \times 9 \div 1$

9. $\dfrac{1}{6} \div \dfrac{3}{8} - \dfrac{4}{9} + \dfrac{5}{8} \times \dfrac{1}{4}$

10. $1.6 \times 0.2 \div 2 + 1.1$

11. $7 \times 3 + 8 \div 4 - 2 - 5$

12. $58 + 7 - 34 + 71$

Answers

1. 1   2. 6   3. 32   4. 18.8   5. 7

6. $\dfrac{1}{10}$ 7. 11   8. 55   9. $\dfrac{5}{32}$ 10. 1.26

11. 16   12. 102

Exponents

An EXPONENT is a small number that is positioned just to the right of and slightly above another number (called the BASE). The numbers in color below are all exponents.

$$16^2 \qquad 8^4 \qquad 9^2 \qquad 10^{10} \qquad 13^5 \qquad 7^3$$

An exponent is a short way of writing a certain kind of multiplication.

Example 3^2 This is read "three to the second power" or "three squared."

$$3^2 = 3 \times 3 = 9 \quad (\text{NOT } 3 \times 2)$$

An exponent of 2 means that the base is multiplied by itself. (The base is used as a factor twice.)

Another Example 3^3 This is read "three to the third power" or "three cubed."

$$3^3 = 3 \times 3 \times 3 = 27$$

An exponent of 3 means that the base is multiplied by itself and then multiplied by itself again. (The base is used as a factor 3 times.)

Another Example 3^4 This is read "three to the fourth power."

$$3^4 = 3 \times 3 \times 3 \times 3 = 81$$

An exponent of 4 means that the base is used as a factor 4 times.

Another Example 3^5 This is read "three to the fifth power."

$$3^5 = 3 \times 3 \times 3 \times 3 \times 3 = 243$$

An exponent of 5 means that the base is used as a factor 5 times.

Exponents are not new to you. You used them in the units on geometry. Area was measured in cm^2, $inches^2$, and so on. Volume was measured in cm^3, $inches^3$, and so on.

Exercises on exponents are on the next page.

Exercises

For each of the following, write out (FACTOR) the expression and give the number that is equal to the expression. For instance: $5^4 = 5 \times 5 \times 5 \times 5 = 625$.

1. 4^2 **2.** 2^5 **3.** 3^6

4. 2^2 **5.** 4^3 **6.** 6^2

7. 5^3 **8.** 4^4 **9.** 8^2

10. 2^6 **11.** 10^2 **12.** 10^3

13. 10^4 **14.** 7^2 **15.** 11^2

16. 7^4 **17.** 1^3 **18.** 1^2

19. 4^6 **20.** 2^7 **21.** 13^2

Answers

1. $4^2 = 4 \times 4 = 16$
2. $2^5 = 2 \times 2 \times 2 \times 2 \times 2 = 32$
3. $3^6 = 3 \times 3 \times 3 \times 3 \times 3 \times 3 = 729$
4. $2^2 = 2 \times 2 = 4$
5. $4^3 = 4 \times 4 \times 4 = 64$
6. $6^2 = 6 \times 6 = 36$
7. $5^3 = 5 \times 5 \times 5 = 125$
8. $4^4 = 4 \times 4 \times 4 \times 4 = 256$
9. $8^2 = 8 \times 8 = 64$
10. $2^6 = 2 \times 2 \times 2 \times 2 \times 2 \times 2 = 64$
11. $10^2 = 10 \times 10 = 100$
12. $10^3 = 10 \times 10 \times 10 = 1000$
13. $10^4 = 10 \times 10 \times 10 \times 10 = 10{,}000$
14. $7^2 = 7 \times 7 = 49$
15. $11^2 = 11 \times 11 = 121$
16. $7^4 = 7 \times 7 \times 7 \times 7 = 2401$
17. $1^3 = 1 \times 1 \times 1 = 1$
18. $1^2 = 1 \times 1 = 1$
19. $4^6 = 4 \times 4 \times 4 \times 4 \times 4 \times 4 = 4096$
20. $2^7 = 2 \times 2 \times 2 \times 2 \times 2 \times 2 \times 2 = 128$
21. $13^2 = 13 \times 13 = 169$

Exponents and Order of Operations

How would you solve something like this?

$$6^2 \div 3^2 - 4^2 \div 2^2$$

In an expression with exponents as well as addition, subtraction, multiplication, and/or division, do the exponents first, then do the rest of the problem.

Example $7^2 \times 2^2 - 4^3 \times 2 - 5^2$

$7^2 \times 2^2 - 4^3 \times 2 - 5^2$ Do all the exponents first.

$49 \times 4 - 64 \times 2 - 25$ Now use the rules for order of operations.

$196 \quad - \quad 128 \quad - 25$

$68 - 25$

43

Another Example $28 \div 2^2 + 16 \times 2^3 - 5^3 - 10$

$28 \div 2^2 + 16 \times 2^3 - 5^3 - 10$ Do all the exponents first.

$28 \div 4 + 16 \times 8 - 125 - 10$ Now use the rules for order of operations.

$7 \quad + 128 \quad - 125 - 10$

$135 \quad - 125 - 10$

$10 - 10$

0

Another Example $3^3 + 2^3 \times 2^2 - 4^2 + 5^2$

$3^3 + 2^3 \times 2^2 - 4^2 + 5^2$

$27 + 8 \times 4 - 16 + 25$

$27 + 32 - 16 + 25$

$59 - 16 + 25$

$43 + 25$

68

Exercises

Work each of the following problems.

1. $2^3 + 4^2 \div 2 \times 2^2$

2. $16 + 40 \div 2^2 + 10^2 \div 5^2$

3. $3^3 + 2^3 \times 2^2$

4. $4^2 \times 3^2 - 6^2 \times 2$

5. $4^3 - 5 \times 3^2 - 2 + 3^3$

6. $7^2 \div 7 + 6^2 \div 3 - 2^2$

7. $8^3 \div 2^2 + 4^3 + 7^2$

8. $4 + 3^3 \times 2^4 + 3^4 \div 9$

9. $12^2 \times 2^3 \div 3 + 4^2 \times 3^2$

10. $7^2 + 6^2 \div 2 - 8^2 + 16$

11. $5^3 \div 25 + 3^2 + 80$

12. $40^2 - 10^2 - 1$

Answers

1. 40	**2.** 30	**3.** 59	**4.** 72	**5.** 44
6. 15	**7.** 241	**8.** 445	**9.** 528	**10.** 19
11. 94	**12.** 1499			

Parentheses and Order of Operations

Sometimes you'll see one or more sets of parentheses or brackets in an arithmetic expression.

$$9 + 5 \times (2 + 3)$$

Parentheses used like this mean that you are to do the operation inside the parentheses before you do any of the rest of the problem.

Example $2 + 24 \div (7 + 5)$

$2 + 24 \div \underbrace{(7 + 5)}$ Do whatever is inside the parentheses first.

$2 + \underbrace{24 \div 12}$ Now use the rules for order of operations.

$\underbrace{2 + 2}$

4

Another Example $(8 + 9) \times 3 \times (4 \div 2)$

$\underbrace{(8 + 9)} \times 3 \times \underbrace{(4 \div 2)}$ Do what is inside the parentheses first.

$\underbrace{17 \times 3} \times 2$ Now use the rules for order of operations.

$\underbrace{51 \times 2}$

102

Another Example $3.1 \times \underbrace{(4.2 + 2.1)} \div 3 + \underbrace{(9.1 - 0.27)} \times 0.4$

$\underbrace{3.1 \times 6.3} \div 3 + 8.83 \times 0.4$

$\underbrace{19.53 \div 3} + 8.83 \times 0.4$

$6.51 + \underbrace{8.83 \times 0.4}$

$\underbrace{6.51 + 3.532}$

10.042

To see the effect that parentheses have on expressions, work the above examples as if there were no parentheses. The answers you get will be quite different.

More exercises are next.

Exercises

1. $44 - 2 \times (8 + 4)$

2. $6 \times (5 - 3)$

3. $(12 - 10) \times 9 + 8 \div (7 - 5)$

4. $8 \div (3 + 1) \times (14 - 4)$

5. $99 + 60 \div (8 - 5) + 14$

6. $(41 - 12) \times 2 - (7 + 2)$

7. $88 \div (31 + 13) - (3 - 1)$

8. $100 - 66 \div (4 - 1)$

9. $11 \times (13 - 8)$

10. $1.7 \times (0.1 + 0.4)$

11. $(2.4 + 3.1) \times (5.4 - 1.6)$

12. $46.8 \div (3.2 - 0.2) \times 0.7$

Answers

1. 20 **2.** 12 **3.** 22 **4.** 20 **5.** 133

6. 49 **7.** 0 **8.** 78 **9.** 55 **10.** 0.85

11. 20.9 **12.** 10.92

Parentheses and Exponents

Don't be surprised when you see an exponent and parentheses in the same expression.

$$(4 + 5)^2$$

This exponent, placed *outside* the parentheses, means that $(4 + 5)$ is multiplied by itself. First add $4 + 5$, then do the exponent.

$$(4 + 5)^2 = (9)^2$$

$$= 9 \times 9 = 81$$

When you see an exponent outside parentheses, do what is inside the parentheses first, then do the exponent.

Example $4 \times (6 - 3)^2$

$4 \times \underbrace{(6 - 3)^2}$ First do the operation inside the parentheses.

$4 \times \quad 3^2$ Now do the exponent.

4×9 Then multiply.

$\underbrace{\qquad}$
36

Another Example $2^3 + (8 - 2)^2 \times (4 - 1)^2$

$2^3 + \underbrace{(8 - 2)^2} \times \underbrace{(4 - 1)^2}$ Do what is inside the parentheses first.

$2^3 + 6^2 \quad \times \quad 3^2$ You know how to work this. Just do the exponents first, then use the rules for order of operations.

$8 + 36 \times 9$

$\underbrace{8 + 324}$

332

Another Example $\left(\dfrac{1}{2}\right)^3 + \left(\dfrac{1}{2} - \dfrac{3}{8}\right) - \left(\dfrac{1}{4}\right)^2$

$\left(\dfrac{1}{2} \times \dfrac{1}{2} \times \dfrac{1}{2}\right) + \left(\dfrac{4}{8} - \dfrac{3}{8}\right) - \left(\dfrac{1}{4} \times \dfrac{1}{4}\right)$

$\underbrace{\dfrac{1}{8}} + \underbrace{\dfrac{1}{8}} - \underbrace{\dfrac{1}{16}}$

$\underbrace{\dfrac{2}{8}} - \dfrac{1}{16}$

$\underbrace{\dfrac{4}{16} - \dfrac{1}{16}}$

$\dfrac{3}{16}$

Exercises

1. $(42 - 34)^2 + 10^2$

2. $(7 + 8) - (3 - 1)^3$

3. $16 + (2 \times 3)^2 + (51 - 49)^3$

4. $(7.6 - 4.3)^2 + 2^4 + 5^2$

5. $32 \times (14 - 12)^2 \div 2^3$

6. $\left(\frac{1}{3}\right)^2 + \left(\frac{1}{3} + \frac{1}{6}\right)^3 - \left(\frac{1}{2}\right)^4$

7. $16 \div 4 + (55 \div 5)^2$

8. $3^3 + 2^2 + (49 - 41)^2$

9. $100^2 + (18 \div 6)^3 + \left(1\frac{1}{2} + \frac{1}{2}\right)^4$

10. $(7 - 0)^2 - (16 - 13)^3$

11. $\left(\frac{1}{2} + \frac{1}{4}\right) \div \left(\frac{1}{2} - \frac{1}{4}\right) + \left(\frac{1}{2}\right)^2$

12. $13 + (50 - 45)^2 \times 3$

Answers

1. 164   **2.** 7   **3.** 60   **4.** 51.89   **5.** 16

6. $\frac{25}{144}$ **7.** 125   **8.** 95   **9.** 10,043   **10.** 22

11. $3\frac{1}{4}$ **12.** 88

We can get different answers using the same numbers, just by moving the exponent around.

$$(2 + 3)^2 \qquad\qquad 2 + 3^2 \qquad\qquad 2^2 + 3$$

$$5^2 \qquad\qquad\qquad 2 + 9 \qquad\qquad 4 + 3$$

$$25 \qquad\qquad\qquad\quad 11 \qquad\qquad\quad 7$$

Now take a look at exponents inside parentheses.

$$(2 + 5^2)$$

When an exponent is *inside* parentheses, it refers *only* to the number it is next to.

$$(2 + 5^2) = (2 + 5 \times 5) \qquad \text{The 5 is multiplied by itself.}$$

$$= (2 + 25)$$

$$= 27$$

Example $(7 + 4^3) - 44$

$(7 + 4^3) - 44$ The exponent refers only to the 4.

$(7 + 64) - 44$ Do the exponent first.

$71 - 44$ Then do what is inside the parentheses.

27

Sometimes you will have exponents inside and outside parentheses in the same problem. When working such problems, ALWAYS BEGIN INSIDE THE PARENTHESES AND WORK OUT.

Example $(11 - 3^2)^4$

$(11 - 3^2)^4$ Start inside the parentheses.
          The exponent 2 refers only to the 3. Do it first.
$(11 - 9)^4$ Next subtract $11 - 9$.

2^4 Now do the outside exponent 4.

16

More exercises are next.

Exercises

1. $\left(\dfrac{1}{4} + 2^2\right) \div \dfrac{1}{8}$

2. $16 - 4 + (5^3 - 10)$

3. $(4^2 - 9) \times 3 + 4^2$

4. $42 - 7 \times (18 - 2^4)$

5. $3 \times (1 + 5^2) \times 2$

6. $4 \times (3^2 - 2^2) \div 2$

7. $20 - (40 - 3^3)$

8. $400 - (130 - 4^3) \times 2$

9. $33 + (11 + 3^2) \times 4$

In Problems 10–15 remember to begin inside the parentheses and work out.

10. $(39 - 6^2)^3$

11. $(7^2 - 3^3)^2$

12. $(5 + 2^2)^2$

13. $10 \times (3^2 - 2^3)^3$

14. $8 \times (3^3 - 5^2)^2$

15. $(21 - 4^2)^3 \div 5$

Answers

1. 34	**2.** 127	**3.** 37	**4.** 28	**5.** 156
6. 10	**7.** 7	**8.** 268	**9.** 113	**10.** 27
11. 484	**12.** 81	**13.** 10	**14.** 32	**15.** 25

Parentheses and Multiplying

Just one more small thing to remember about parentheses. When there is no sign (no $+$, $-$, $\times$, or $\div$) you MULTIPLY.

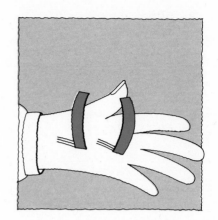

Example   2(4)

⟵ There is no sign, so you multiply.

2(4) = 2 × 4 = 8

Another Example $(4)(5) + (3)(4)^2$

No sign means you multiply.

$(4)(5) + (3)(4)^2$ The exponent refers just to what is in the nearest
  ↓   ↓   parentheses—just the 4, not the 3.
$4 \times 5 + 3 \times 16$ $(4)^2 = 4 \times 4 = 16$

  20   +   48

        68

Note that the rules for order of operations stay the same.

Another Example $3(4 + 2)^2$

$3(4 + 2)^2$ First do what is inside the parentheses.

$3(6)^2$ The exponent refers just to the 6. $(6)^2 = 6 \times 6 = 36$
  ↓
3×36

  108

Why would someone decide that parentheses with no sign means multiplication? In algebra you use the letter "x" a lot. If you also used the times sign (×), you might get the two confused. So in algebra you use parentheses and no sign to show multiplication.

Try some exercises.

Exercises

1. $3(7) + 3(8)$ **2.** $3(5) + 2(4)$ **3.** $4(3) - 2(3)$

4. $(3)(5) - (4)(3)$ **5.** $3(6 - 2)$ **6.** $(5 - 2)(2 + 3)$

7. $6(5 - 1) - 3(5 - 1)$ **8.** $3(4 - 1) + 2(6 - 2)(4 - 2)$ **9.** $6(8 - 2) + 3(4)$

Watch carefully for the exponents in the rest of the problems.

10. $4^2 - 2(3)(2)$ **11.** $2(4 - 2 + 3)^2$ **12.** $(10 - 2)^2 - (8 - 3 - 3)^2$

13. $3(2 + 2)^2$ **14.** $5(3 - 1)^3$ **15.** $2(4 - 1)^3 - 3(4 + 1)^2 \div 5$

Answers

15. 39	**14.** 40	**13.** 48	**12.** 60	**11.** 50
10. 4	**9.** 48	**8.** 25	**7.** 12	**6.** 15
5. 12	**4.** 3	**3.** 6	**2.** 23	**1.** 45

Involved Fractions

What would you do if you came across something like this?

$$\frac{3 + 4(2) - 1}{2(3) + 2 - 3}$$

Don't leave! This is just a fraction—an involved one, but still a fraction. If you solve the top, and then solve the bottom, you will end up with an easy division problem.

Here the top is worked up,

$$10$$
$$11 - 1$$
$$3 + 8 - 1$$
$$3 + 4 \times 2 - 1$$

$$\frac{3 + 4(2) - 1}{2(3) + 2 - 3} = \frac{10}{5} = 10 \div 5 = 2$$

and the bottom is worked down.

$$2 \times 3 + 2 - 3$$
$$6 + 2 - 3$$
$$8 - 3$$
$$5$$

You end up with 10 on the top and 5 on the bottom. The final answer is 2.

Another Example

$$\frac{(5)(2) - 8 \div 2}{7\frac{1}{2} - 3\frac{1}{2}}$$

$$6$$
$$10 - 4$$
$$5 \times 2 - 8 \div 2$$

$$\frac{(5)(2) - 8 \div 2}{7\frac{1}{2} - 3\frac{1}{2}} = \frac{6}{4} = 1\frac{2}{4} = 1\frac{1}{2}$$

$$4$$

More exercises are next.

Exercises

1. $\dfrac{8 \div 2 + 3(2)}{5 + 1}$

2. $\dfrac{5 + 2(3) - 1}{12 \div 2 - 1}$

3. $\dfrac{18 - (2)(3) - 4 - (2)(3)}{19 - 9}$

4. $\dfrac{10 \div 2 - 4 \div 2 + 3}{6(2) - 12 \div 4}$

5. $\dfrac{6(2) - 8 \div 4}{5}$

6. $\dfrac{6(10) \div 5 - 4(2)}{20 - 4(5 - 3)}$

7. $\dfrac{2(6 + 8 \div 4)(6 - 2)}{(6 - 4 \div 2)^2}$

8. $\dfrac{\dfrac{1}{2}}{\dfrac{1}{4}}$ ← This line means "divided by."

$\dfrac{1}{2} \div \dfrac{1}{4}$

9. $\dfrac{\dfrac{1}{4} + \dfrac{1}{2}}{\dfrac{3}{4} - \dfrac{1}{2}}$

10. $\dfrac{\dfrac{2}{3} \times \dfrac{3}{4}}{2}$

11. $\dfrac{\dfrac{1}{2} + \dfrac{1}{3}}{\dfrac{1}{6}}$

12. $\dfrac{\dfrac{1}{5}}{\dfrac{1}{10}}$

13. $\dfrac{\dfrac{3}{4}}{\dfrac{1}{8}}$

14. $\dfrac{\dfrac{3}{8} - \dfrac{1}{4}}{\left(\dfrac{1}{2}\right)\left(\dfrac{1}{4}\right)}$

15. $\dfrac{\left(\dfrac{1}{4}\right)^2}{\left(\dfrac{1}{2}\right)^3}$

Answers

1. $1\dfrac{2}{3}$ **2.** 2   **3.** $\dfrac{1}{5}$ **4.** $\dfrac{2}{3}$ **5.** 2

6. $\dfrac{1}{3}$ **7.** 4   **8.** 2   **9.** 3   **10.** $\dfrac{1}{4}$

11. 5   **12.** 2   **13.** 6   **14.** 1   **15.** $\dfrac{1}{2}$

Time for a review. These exercises cover everything in this unit.

Review Exercises

The first ones are easy. Just remember the rules for order of operations.

1. $3 \times 2 - 4 + 2$

2. $8 \div 2 + 1 \times 3 - 3$

3. $15 - 2^2 \times 3$

4. $4 - 2 \div 2 + 3 \times 2 - 4 \div 2$

5. $4^2 \times 3^2$

6. $30 + 0.84 \div 2 - 15.21 \times 2$

Now you need to pay close attention to the parentheses and where the exponents are placed. (Notice that Problems 7–12 all use the same three numbers, but have different answers. It's very important to do things in the right order!)

7. $4(3 + 2)^2$

8. $4(3^2 + 2^2)$

9. $4^2(3 + 2)^2$

10. $4(3) + 2^2$

11. $4^2(3)^2 + 2^2$

12. $4(3)^2 + 2$

Answers

1. 4	2. 4	3. 3	4. 7	5. 144
6. 0	7. 100	8. 52	9. 400	10. 16
11. 148	12. 38			

13. $4(4-2)^3$ **14.** $3(2+6)^2$ **15.** $(4-2)(4^2-3^2)$

16. $3(6^2-18)+2(4^2-2^2)$ **17.** $(6-4)^2(4+8\div 2)^2$ **18.** $\dfrac{1}{2}-\left(\dfrac{1}{4}\right)^2+3(2^2+1)$

Solve the top part of the fraction first, then the bottom part. Simplify each answer.

19. $\dfrac{(7-4)^3}{3^2}$ **20.** $\dfrac{(10-8)^2}{(4+14\div 7)^2}$ **21.** $\dfrac{\left(\dfrac{1}{2}\right)^3}{\left(\dfrac{1}{4}\right)^2}$

22. $\dfrac{\left(\dfrac{1}{3}-\dfrac{1}{6}\right)^2}{\left(\dfrac{1}{2}+\dfrac{1}{3}\right)^2}$ **23.** $\dfrac{\dfrac{2}{3}\left(\dfrac{1}{2}\div\dfrac{1}{4}\right)^2}{6\div\left(\dfrac{1}{3}+\dfrac{1}{6}\right)^2}$ **24.** $\dfrac{\left(\dfrac{2}{3}-\dfrac{1}{4}\right)^2}{\left(\dfrac{3}{4}-\dfrac{1}{2}\div 4\right)^2}$

Answers

13. 32	**14.** 192	**15.** 14	**16.** 78
17. 256	**18.** $15\dfrac{7}{16}$	**19.** 3	**20.** $\dfrac{1}{9}$
21. 2	**22.** $\dfrac{1}{25}$	**23.** $\dfrac{1}{9}$	**24.** $\dfrac{4}{9}$

Unit 16 Test

Use a separate piece of paper to work out each problem.

1. $16 \times (2 + 3) - 20$

2. $48 - 36 + 4 \times (1 + 1)$

3. $13 - 3^2 - 2^2$

4. $3 \times (4^2 + 1)$

5. $4 \times (15 - 2^3)^2$

6. $88 + 3 - (4 + 2)^2$

7. $3(41 - 39)^2 + 4^2$

8. $218 - (3^2 - 1)(2 + 2)$

9. $3(4 + 5)^2(3^2 - 4)$

10. $7(6 - 2^2)^3$

11. $13(4^2 - 15) + 3(3)^2$

12. $400 - (10 \div 2)^3(2)$

13. $4(2)^3(83 - 78)^2 + 10$

14. $(4 - 2)^4 + (8 - 6)^4$

15. $100 - (100 - 90)^2$

16. $(2)^2(3)^3 + (2)^3(3)^2$

17. $\dfrac{(40 \div 4)^2 - 75}{4 + 1}$

18. $\dfrac{(16 \div 2)^2 + (4^2 - 3)}{(14 - 6)^2 - 53}$

19. $\dfrac{(71 - 65)^2 - 10}{3^3 + 25}$

20. $\dfrac{4 + 3(2 - 1) + 2^3}{(4)^2 - 1}$

21. $\dfrac{\left(\dfrac{1}{4}\right)^2 + \dfrac{5}{16}}{2}$

22. $\dfrac{\left(1 + \dfrac{1}{2}\right)^2 + \dfrac{1}{4}}{\left(\dfrac{1}{2}\right)^2}$

Your Answers

1. _____
2. _____
3. _____
4. _____
5. _____
6. _____
7. _____
8. _____
9. _____
10. _____
11. _____
12. _____
13. _____
14. _____
15. _____
16. _____
17. _____
18. _____
19. _____
20. _____
21. _____
22. _____

17 | Preparing for Algebra: Equations

Aims you toward:

- Evaluating expressions with letters and numbers.
- Solving equations by adding the same number to both sides.
- Solving equations by subtracting the same number from both sides.
- Solving equations by multiplying both sides by the same number.
- Solving equations by dividing both sides by the same number.
- Solving two-step equations.

In this unit you will work exercises like these:

- What is the value of $3 + a + bc$ if $a = 4$, $b = 2$, and $c = 9$.
- Find the value of x in this equation: $x - 10 = 5^2$.
- Find the value of f in this equation: $3(8) = f + 2$.
- Find the value of z in this equation: $\dfrac{z}{3} = 4$.
- Find the value of r in this equation: $40 + 2 = 7r$.
- Find the value of s in this equation: $9s + 6 = 42$.

Equations

Algebra deals mainly with equations. You can recognize an equation by its = sign. The part of an equation to the left of the = sign is called the LEFT SIDE. The part to the right of the = sign is called the RIGHT SIDE. Remember that in an equation the left side is always equal to the right side.

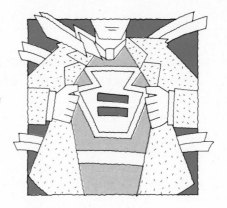

Example $7 - 4 = 2 + 1$

   $3 \quad = \quad 3$ This is an equation because the left side ($7 - 4$) is equal to the right side ($2 + 1$).

Another Example $4 \times 6 \neq 3 \times 7$

   $24 \quad \neq \quad 21$ This is *not* an equation because the left side equals 24 and the right side equals 21.

You use the $\neq$ ("does not equal") sign to show that you do *not* have an equation.

Sometimes you will have two expressions and you will need to decide whether or not they make an equation.

Example   Put either an = sign or a $\neq$ sign in the box between the two expressions.

   $4(2) + 3 \ \boxed{} \ 2 + 10 - 1$ Solve each expression.

   $8 + 3 \qquad 12 - 1$

   $11 \quad = \quad 11$ Both sides are the same, so put an = sign in the box. You have an equation.

Another Example   Put either an = sign or a $\neq$ sign in the box between the two expressions.

   $\dfrac{48}{3} \ \boxed{} \ 5(3)$ Solve each expression.

   $16 \ \neq \ 15$ The sides are not the same, so put a $\neq$ sign in the box. You do *not* have an equation.

The exercises on the next page will give you more practice with equations.

Exercises

For each of the following, put either an $=$ ("equals") sign or a $\neq$ ("does not equal") sign in the box between the two expressions.

1. $4(3)^2 \;\square\; 36$

2. $13 - 1 \;\square\; \dfrac{6^2}{3}$

3. $9^2 - 17 \;\square\; 2(30)$

4. $23 + 35 \;\square\; 10^2 - 42$

5. $(9 - 8)3 \;\square\; 6 - 2^2$

6. $340 - 30 \;\square\; 3(10^2) + 2(5)$

7. $500 - 2(250) \;\square\; 19 - 19$

8. $(11)(16) \;\square\; (10)(17)$

9. $\left(\dfrac{1}{2}\right)\left(\dfrac{3}{4}\right) \;\square\; 7 - 6\dfrac{5}{8}$

10. $(4.5)2 \;\square\; 2^3$

11. $\dfrac{42}{7}(2) \;\square\; \dfrac{33}{3}$

12. $63 - 4^2 \;\square\; (2)(3)(7)$

13. $19 + 3\dfrac{1}{2} \;\square\; \dfrac{30 + 15}{2}$

14. $12 - 8 \;\square\; 15 - 10$

15. $\dfrac{(4)(8)}{2} \;\square\; 4^2$

16. $\dfrac{3}{8} + \dfrac{1}{2} \;\square\; 1\dfrac{5}{8} - \dfrac{3}{4}$

Answers

13. $=$	**14.** $\neq$	**15.** $=$	**16.** $=$
9. $=$	**10.** $\neq$	**11.** $\neq$	**12.** $\neq$
5. $\neq$	**6.** $=$	**7.** $=$	**8.** $\neq$
1. $=$	**2.** $=$	**3.** $\neq$	**4.** $=$

Letters, Numbers, and Equations

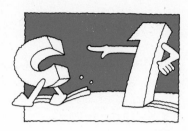

In algebra numbers are often represented by letters. To find the value of an expression that has letters, you replace the letters with their number values.

Example   What is the value of $(c + 4 + d)3$ if $c = 1$ and $d = 4$?

$(c + 4 + d)3$ Replace the letters with their number values.

$(1 + 4 + 4)3$ Work the problem as usual.

$9(3)$

27

When there is no sign (no $+$, $-$, $\times$, or $\div$) between a number and a letter, or between 2 letters, you MULTIPLY.

Example   What is the value of $5a + bc$ if $a = 3$, $b = 12$, and $c = 2$?

$5a + bc$

$5(a) + (b)(c)$ Replace the letters with their number values and

$5(3) + (12)(2)$ work the problem as usual.

$15 + 24$

39

Here's an example of how to tell if two expressions with letters make an equation.

Example   Put either an $=$ sign or a $\neq$ sign in the box between the two expressions.

Assume that $x = 25$, $b = 3$, and $c = 12$.

$60 - x + 3b \ \square\ 3c + b + 5$ Replace the letters in each expression with their number

$60 - 25 + 3(3) \quad 3(12) + 3 + 5$ values. Solve each expression.

$60 - 25 + 9 \qquad 36 + 3 + 5$

$35 + 9 \qquad\quad 39 + 5$

$44 \quad = \quad 44$ Both sides equal 44, so put an $=$ sign in the box. You have an equation.

Exercises

Find the value of each of the following expressions. Assume that $a = 4$, $b = 7$, $c = 5$, $x = 3$, and $y = 9$.

1. $x + 9$

2. $8c + 3a + 5$

3. $7b$

4. $2(2y - x)$

5. $2a + 8$

6. $\dfrac{4a + 3x - 1}{a}$

For each of the following, indicate whether or not the two expressions make an equation. Put either an $=$ sign or a $\neq$ sign in each of the boxes. Assume that $b = 8$, $d = 4$, $x = 3$, $y = 7$, and $z = 10$.

7. $d - x \;\boxed{}\; z - y - 2$

8. $4z + 2b + 2 \;\boxed{}\; by$

9. $3(y - 2) \;\boxed{}\; z + x + 2$

10. $6 + 8z - 9 \;\boxed{}\; b^2 + 3d + 1$

11. $5b + 2 \;\boxed{}\; 6y$

12. $d^2 + 2 \;\boxed{}\; x^3$

13. $\dfrac{5x - 3d + 1}{d} \;\boxed{}\; d - x$

14. $8(5b - 9d) \;\boxed{}\; x + z$

15. $\dfrac{4y + 2}{6} \;\boxed{}\; y - 2$

Answers				
1. 12	2. 57	3. 49	4. 30	5. 16
6. 6	7. $=$	8. $\neq$	9. $=$	10. $=$
11. $=$	12. $\neq$	13. $=$	14. $\neq$	15. $=$

Solving Equations

So far when you have had equations with letters in them you have always known the number value of the letters. Now you will begin to see equations with an unknown letter.

Look at this equation, for example:

$$x + 2 = 5.$$

You don't know the value of x. All you know is that $x + 2$ is equal to 5. If you think about it for a minute, you realize that x must equal 3. If you replace the x with a 3, the equation makes a true statement.

$$x + 2 = 5$$
$$3 + 2 = 5$$
$$5 = 5 \quad \text{Solution: } x = 3$$

That's all there is to solving an equation. You just find the number value of the unknown letter.

Another Example   What is the value of a in this equation: $a - 1 = 7$?

You know that $8 - 1 = 7$, so a must equal 8.

$$a - 1 = 7$$
$$8 - 1 = 7$$
$$7 \quad \text{Solution: } a = 8$$

Another Example   What is the value of r in this equation: $r - 4 = 10$?

You know that $14 - 4 = 10$, so r must equal 14.

$$r - 4 = 10$$
$$14 - 4 = 10$$
$$10 = 10 \quad \text{Solution: } r = 14$$

You can't solve all equations by just looking at them. To solve equations you sometimes need to add, subtract, multiply, or divide. But you can't do the operations any way you like. There's one basic rule to remember about solving equations:

Whatever you do to one side you must also do to the other.

Think of an equation as a balance. The left side equals the right side.

You can add the same number to both sides and still keep the balance.

You can subtract the same number from both sides and still keep the balance.

You can multiply both sides by the same number and still keep the balance.

You can divide both sides by the same number and still keep the balance.

But if you ever do something to one side and not the other you lose the balance and no longer have an equation.

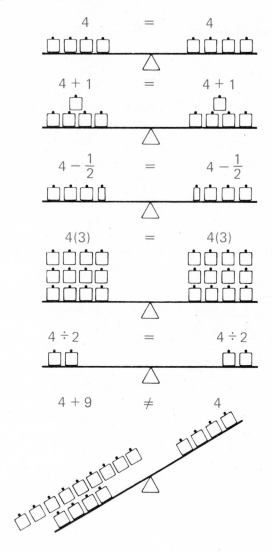

Along with keeping the two sides balanced, there's a little trick to solving equations:

Work with the equation until you get the unknown letter alone on one side of the equation.

Keep this trick in mind. Its purpose will become clearer as you go on . . . .

Solving Equations by Adding

There is one type of equation that can be solved by adding the same number to both sides.

First, here's a reminder about addition and subtraction:

> Addition and subtraction are the opposite operations of each other. If you first add and then subtract (or subtract and then add) the same number, you haven't changed anything.

Examples
$$34 - 7 + 7$$
$$27 + 7$$
$$34$$

$$38 + 400 - 38$$
$$438 - 38$$
$$400$$

$$x - 8 + 8$$
$$x$$

Back to solving equations by adding . . .

The first step to solving equations is to get the unknown letter alone on one side. In equations like the ones in the following examples, where a number is subtracted from the unknown letter, you get the unknown letter alone by adding.

Example   Solve this equation: $x - 9 = 13$. Check the solution.

$$x - 9 = 13$$ You want to get the x alone on the left side. Observe: The equation has $x - 9$. Ask: How can I get rid of the 9?

$$x - 9 + 9 = 13 + 9$$ Add 9 to both sides.

$$x - 9 + 9 = 13 + 9$$ Simplify the left side. Subtracting 9 and adding 9 are opposites, so you just have x on the left

$$x = 13 + 9$$ side.

$$x = 22$$ Now simplify the right side.

To check the solution, go to the original equation and replace x with 22.

$$x - 9 = 13$$
$$22 - 9 = 13$$
$$13 = 13$$

The solution, $x = 22$, is correct.

Another Example

Solve this equation: $q - 5 = 24$. Check the solution.

$$q - 5 = 24$$ You want the q alone on the left side.
                  Observe: The equation has $q - 5$.
                  Ask: How can I get rid of the 5?

$$q - 5 + 5 = 24 + 5$$ Add 5 to both sides.

$$q = 24 + 5$$ Simplify the left side.

$$q = 29$$ Simplify the right side.

Check: $q - 5 = 24$

$$29 - 5 = 24$$

$$24 = 24$$

The solution, $q = 29$, is correct.

Another Example

What is the value of b in this equation: $b - 400 = 30 + 7$?

$$b - 400 = 30 + 7$$ First add $30 + 7$.

$$b - 400 = 37$$ Now get the b alone on the left side.

$$b - 400 + 400 = 37 + 400$$ Add 400 to both sides.

$$b = 37 + 400$$ Simplify the left side.

$$b = 437$$ Simplify the right side.

Another Example

Solve this equation: $15(2) - 10 = a - 17$.

$$15(2) - 10 = a - 17$$ First do the left side.

$$30 - 10 = a - 17$$

$$20 = a - 17$$ Now get the a alone on the right side.

$$20 + 17 = a - 17 + 17$$ Add 17 to both sides.

$$20 + 17 = a$$ Simplify the right side.

$$37 = a$$ Simplify the left side.
              The solution is $37 = a$, or $a = 37$.

Work the exercises on the next page using the method shown in the examples above.

Exercises

Find the value of the unknown letter in each of the following equations. Check each solution.

1. $m - 3 = 7$

2. $n - 52 = 87$

3. $94 = g - 35$

4. $7 = x - 8$

5. $x - 5 = 2 + 6$

$x - 5 = 8$

6. $k - 8 = 5^2$

7. $d - 43 = 43$

8. $68 = c - 15$

9. $b - 25 = 4^2 - 3$

10. $v - 37 = 10 - 3$

11. $2(30) - 12 = w - 10$

12. $6^2 - 5 = y - 59$

13. $d - 66 = (10)2 + 4$

14. $14 = f - 71$

15. $z - 50 = 9^2 + 2(3)$

Answers

1. $m = 10$ **2.** $n = 139$ **3.** $g = 129$ **4.** $x = 15$ **5.** $x = 13$

6. $k = 33$ **7.** $d = 86$ **8.** $c = 83$ **9.** $b = 38$ **10.** $v = 44$

11. $w = 58$ **12.** $y = 90$ **13.** $d = 90$ **14.** $f = 85$ **15.** $z = 137$

Solving Equations by Subtracting

The equations on the last two pages were solved by adding the same number to both sides. Some equations are solved by subtracting the same number from both sides.

Example   Solve this equation: $c + 9 = 40$. Check the solution.

$$c + 9 = 40$$ You want the c alone on the left side.
Observe: The equation has $c + 9$.
Ask: How can I get rid of the 9?

$$c + 9 - 9 = 40 - 9$$ Subtract 9 from both sides.

$$c = 40 - 9$$ Simplify the left side.

$$c = 31$$ Simplify the right side.

Check: $c + 9 = 40$

$$31 + 9 = 40$$

$$40 = 40$$

The solution, $c = 31$, is correct.

In equations where a number is added to the unknown letter, you get the unknown letter alone by subtracting.

Another Example   What is the value of y in $7 + y = 2(24)$?

$$7 + y = 2(24)$$ First multiply 2(24).

$$7 + y = 48$$ You want y alone on the left side.

$$7 + y - 7 = 48 - 7$$ Subtract 7 from both sides.

$$y = 48 - 7$$ Simplify the left side.

$$y = 41$$ Simplify the right side.

Another Example   What is the value of k in $3^2 = k + 4$?

$$3^2 = k + 4$$ First do 3^2.

$$9 = k + 4$$ You want the k alone on the right side.

$$9 - 4 = k + 4 - 4$$ Subtract 4 from both sides.

$$9 - 4 = k$$ Simplify the right side.

$$5 = k$$ Simplify the left side.

More exercises are next.

Exercises

Find the value of the unknown letter in each of the following equations. Check each solution.

1. $x + 5 = 13$

2. $m + 6 = 9$

3. $3 + x = 10$

4. $b + 18 = 49 + 4$
$b + 18 = 53$

5. $245 + a = 5(113)$

6. $23 + n = \dfrac{90}{2}$

7. $54 = d + 29$

8. $104 = 82 + b$

9. $q + 35 = 40(2) + 16$

10. $c + 19 = 6^2 + 4$

11. $71 + t = 100 - 17$

12. $g + 25 = 30(3)$

13. $18 + y = 7^2 - 13$

14. $83 - 5 = h + 56$

15. $(41)3 - 21 = 28 + v$

Answers

1. $x = 8$
2. $m = 3$
3. $x = 7$
4. $b = 35$
5. $a = 320$
6. $n = 22$
7. $d = 25$
8. $b = 22$
9. $q = 61$
10. $c = 21$
11. $t = 12$
12. $g = 65$
13. $y = 18$
14. $h = 22$
15. $v = 74$

Solving Equations by Multiplying

You now know how addition and subtraction can be used to solve equations. This page explains how multiplication can be used to solve equations.

First another reminder, this time about multiplication and division:

> Multiplication and division are the opposite operations of each other. If you first multiply and then divide (or divide and then multiply) by the same number, you haven't changed anything.

Examples

$$4 \times 3 \div 3 \qquad 18 \div 6 \times 6 \qquad a \div 3 \times 3$$
$$12 \div 3 \qquad\qquad 3 \times 6 \qquad\qquad a$$
$$4 \qquad\qquad\qquad 18$$

Below are the same examples with multiplication shown by no sign and parentheses and with division shown by a fraction line. This is how you'll see multiplication and division in the rest of the unit.

Examples

$$\frac{4(3)}{3} \qquad\qquad \frac{18}{6}(6) \qquad\qquad \frac{a}{3}(3)$$
$$\frac{12}{3} \qquad\qquad 3(6) \qquad\qquad a$$
$$4 \qquad\qquad\qquad 18$$

The following examples show how some equations can be solved by multiplying both sides by the same number.

Example   Solve this equation: $\dfrac{m}{2} = 8$.

$\dfrac{m}{2} = 8$ You want m on the left side.

Observe: The equation has $\dfrac{m}{2}$.

Ask: How can I get rid of the 2?

$\dfrac{m}{2}(2) = 8(2)$ Multiply both sides by 2.

$m = 8(2)$ Simplify the left side.

$m = 16$ Simplify the right side.

Since multiplication is the opposite operation of division, in equations where the unknown letter is divided by a number you multiply to get the unknown letter alone.

Another Example   Find the value of p in this equation: $\frac{p}{5} = 10$. Check the solution.

$\frac{p}{5} = 10$ You want p alone on the left side.

$\frac{p}{5}(5) = 10(5)$ Multiply both sides by 5.

$p = 10(5)$ Simplify the left side.

$p = 50$ Simplify the right side.

Check: $\frac{p}{5} = 10$

$\frac{50}{5} = 10$

$10 = 10$

The solution, $p = 50$, is correct.

Another Example   Find the value of y in this equation: $7(3) = \frac{y}{4}$.

$7(3) = \frac{y}{4}$ First multiply $7(3)$.

$21 = \frac{y}{4}$ You want y alone on the right side.

$21(4) = \frac{y}{4}(4)$ Multiply both sides by 4.

$21(4) = y$ Simplify the right side.

$84 = y$ Simplify the left side.

Another Example   Solve this equation: $3^2 + 4 = \frac{b}{3}$.

$3^2 + 4 = \frac{b}{3}$ First do the left side.

$9 + 4 = \frac{b}{3}$

$13 = \frac{b}{3}$ You want b alone on the right side.

$13(3) = \frac{b}{3}(3)$ Multiply both sides by 3.

$13(3) = b$ Simplify the right side.

$39 = b$ Simplify the left side.

Exercises

Find the value of the unknown letter in each of the following equations. Check each solution.

1. $\dfrac{x}{21} = 14$

2. $12 = \dfrac{g}{36}$

3. $\dfrac{n}{25} = 3$

4. $\dfrac{e}{4} = 13$

5. $\dfrac{w}{13} = 13$

6. $(5)(4) = \dfrac{s}{4}$

7. $3^2 = \dfrac{y}{200}$

8. $\dfrac{t}{61} = 2^3$

9. $\dfrac{k}{12} = 9$

10. $\dfrac{m}{45} = 2$

11. $\dfrac{m}{8} = 4(40)$

12. $\dfrac{v}{3} = 5^2$

13. $\dfrac{n}{15} = 12 - 7$

14. $\dfrac{d}{7} = 10^2 - 10$

15. $(6)7 = \dfrac{k}{7}$

Answers

1. $x = 294$ **2.** $g = 432$ **3.** $n = 75$ **4.** $e = 52$ **5.** $w = 169$
6. $s = 80$ **7.** $y = 1800$ **8.** $t = 488$ **9.** $k = 108$ **10.** $m = 90$
11. $m = 1280$ **12.** $v = 75$ **13.** $n = 75$ **14.** $d = 630$ **15.** $k = 294$

Solving Equations by Dividing

The examples below show how some equations can be solved by dividing both sides by the same number.

Example   Find the value of z in this equation: $4z = 8$. Check the solution.

$4z = 8$ You want to get z alone on the left side.
Observe: The equation has $4z$.
Ask: How can I get rid of the 4?

$\dfrac{4z}{4} = \dfrac{8}{4}$ Divide both sides by 4.

$z = \dfrac{8}{4}$ Simplify the left side.

$z = 2$ Simplify the right side.

Check: $4z = 8$

$4(2) = 8$

$8 = 8$

The solution, $z = 2$, is correct.

Since division is the opposite operation of multiplication, in equations where the unknown letter is multiplied by a number you divide to get the unknown letter alone.

Another Example   Solve this equation: $8 + 4 = 3z$.

$8 + 4 = 3z$ First add $8 + 4$.

$12 = 3z$ You want z alone on the right side.

$\dfrac{12}{3} = \dfrac{3z}{3}$ Divide both sides by 3.

$\dfrac{12}{3} = z$ Simplify the right side.

$4 = z$ Simplify the left side.

Work the exercises on the next page.

Exercises

Find the value of the unknown letter in each of the following equations. Check each solution.

1. $35n = 140$

2. $126 = 7t$

3. $891 = 33r$

4. $6x = 5 + 3$

5. $6z = 50(3)$

6. $2x = 2$

7. $(50)2 + 8 = 4b$

8. $3a = 2$

9. $12n = (40)3 + (4)3$

10. $3c = 78$

11. $5t = 3(4) + 3$

12. $\dfrac{100 + 152}{2} = 3q$

13. $2(10^2) + 185 = 11y$

14. $207 = 23d$

15. $12v = 5^3 + 67$

Answers

1. $n = 4$
2. $t = 18$
3. $r = 27$
4. $x = 1\dfrac{1}{3}$
5. $z = 25$

6. $x = 1$
7. $b = 27$
8. $a = \dfrac{2}{3}$
9. $n = 11$
10. $c = 26$

11. $t = 3$
12. $q = 42$
13. $y = 35$
14. $d = 9$
15. $v = 16$

Solving Two-Step Equations

In the equations you have solved so far there has been only one step in getting the unknown letter alone on one side. You will now learn how to solve equations that require two steps to get the unknown letter alone on one side.

The examples below show equations where you first add or subtract, and then multiply to get the unknown letter alone.

Example Find the value of t in this equation: $\dfrac{t}{6} + 5 = 8$. Check the solution.

$\dfrac{t}{6} + 5 = 8$ You want to get the t alone on the left side. It will take two steps.

$\dfrac{t}{6} + 5 - 5 = 8 - 5$ First subtract 5 from both sides.

$\dfrac{t}{6} = 3$ This removes the 5 from the left side.

$\dfrac{t}{6}(6) = 3(6)$ Now multiply both sides by 6.

$t = 18$ Simplifying gives you $t = 18$.

Check: $\dfrac{t}{6} + 5 = 8$

$\dfrac{18}{6} + 5 = 8$

$3 + 5 = 8$

$8 = 8$

The solution, $t = 18$, is correct.

Another Example Solve this equation: $\dfrac{y}{8} - 3 = 2$.

$\dfrac{y}{8} - 3 = 2$ It will take two steps to get y alone.

$\dfrac{y}{8} - 3 + 3 = 2 + 3$ First add 3 to both sides.

$\dfrac{y}{8} = 5$ This removes the 3 from the left side.

$\dfrac{y}{8}(8) = 5(8)$ Now multiply both sides by 8.

$y = 40$ Simplifying gives $y = 40$.

The examples below show equations where you first add or subtract, and then divide to get the unknown letter alone.

Example   Find the value of x in this equation: $4x + 12 = 32$.

$$4x + 12 = 32$$ It will take two steps to get the x alone.

$$4x + 12 - 12 = 32 - 12$$ First subtract 12 from both sides.

$$4x = 20$$ This removes the 12 from the left side.

$$\frac{4x}{4} = \frac{20}{4}$$ Now divide both sides by 4.

$$x = 5$$ Simplifying gives $x = 5$.

Another Example   Solve this equation: $180 = 2c - 40$.

$$180 = 2c - 40$$ It will take two steps to get the c alone.

$$180 + 40 = 2c - 40 + 40$$ First add 40 to both sides.

$$220 = 2c$$ This removes the 40 from the right side.

$$\frac{220}{2} = \frac{2c}{2}$$ Now divide both sides by 2.

$$110 = c$$ Simplifying gives $110 = c$, or $c = 110$.

Another Example   Solve this equation: $3x + 1 = 13$.

$$3x + 1 = 13$$ Use two steps to get the x alone.

$$3x + 1 - 1 = 13 - 1$$ First subtract 1 from both sides.

$$3x = 12$$

$$\frac{3x}{3} = \frac{12}{3}$$ Now divide both sides by 3.

$$x = 4$$

To Solve Equations	Here is a general set of rules that always seems to work:
	1. *Add or subtract first,* to get the numbers all on one side of the equals sign.
	2. *Multiply or divide last,* to get the unknown letter by itself.
	3. Always remember to *check your answer* by replacing the unknown letter at the beginning of the problem with your answer.

Example $0.5x - 0.2 = 0.8$

Step 1 $0.5x - 0.2 + 0.2 = 0.8 + 0.2$ Get the numbers on the right side of the equals sign by adding 0.2 to both sides of the equation.

$$0.5x = 1.0$$

Step 2 $$\frac{0.5x}{0.5} = \frac{1.0}{0.5}$$ Get the unknown letter by itself by dividing both sides of the equation by 0.5.

$$x = 2$$

Step 3   To check your answer, replace the unknown letter at the beginning of the example with your answer.

$$0.5x - 0.2 = 0.8$$
$$0.5(2) - 0.2 = 0.8$$
$$1.0 - 0.2 = 0.8$$
$$0.8 = 0.8$$

If the two sides balance, you know that your answer is correct.

Another Example $\dfrac{m}{2} + 3 = 5$

Step 1 $\dfrac{m}{2} + 3 - 3 = 5 - 3$ Subtract 3 from both sides to get the numbers on the right side of the equals sign.

$\dfrac{m}{2} = \dfrac{2}{1}$ A whole number such as 2 can be written as $\dfrac{2}{1}$.

Step 2 $\left(\dfrac{2}{1}\right)\dfrac{m}{2} = \left(\dfrac{2}{1}\right)\dfrac{2}{1}$ Get the unknown letter by itself on the left by multiplying both sides

$m = \dfrac{4}{1}$ or 4   by $\dfrac{2}{1}$.

Step 3   To check your answer, replace the unknown letter at the beginning with your answer.

$\dfrac{m}{2} + 3 = 5$

$\dfrac{4}{2} + 3 = 5$

$2 + 3 = 5$

$5 = 5$ The two sides balance, so your answer is correct.

Exercises

We have room here to try one of each type of problem shown above. Be sure to follow the three steps.

1. $4t - 0.7 = 2.1$ **2.** $\dfrac{p}{3} + 5 = 29$

Answers

Exercises

Find the value of the unknown letter in each of the following equations. Check each solution.

1. $\dfrac{x}{2} + 7 = 11$

2. $\dfrac{n}{6} + 13 = 15$

3. $\dfrac{w}{10} - 1 = 2$

4. $7n + 2 = 51$

5. $4t + 4 = 40$

6. $5a + 8 = 13$

7. $30p + 10 = 340$

8. $\dfrac{c}{3} - 3 = 4$

9. $\dfrac{a}{9} - 6 = 2$

10. $\dfrac{k}{30} + 62 = 65$

11. $1.4k + 0.2 = 3$

12. $\dfrac{b}{1.2} + 4.6 = 11.6$

13. $0.7g + 0.5 = 3.3$

14. $9n - 15 = 93$

15. $\dfrac{h}{0.4} + 2 = 11$

Answers

1. $x = 8$ **2.** $n = 12$ **3.** $w = 30$ **4.** $n = 7$ **5.** $t = 9$
6. $a = 1$ **7.** $p = 11$ **8.** $c = 21$ **9.** $a = 72$ **10.** $k = 90$
11. $k = 2$ **12.** $b = 8.4$ **13.** $g = 4$ **14.** $n = 12$ **15.** $h = 3.6$

The following exercises will give you more practice in solving equations.

Review Exercises

For each of the following, indicate whether or not the two expressions make an equation. Put either an $=$ sign or a $\neq$ sign in each of the boxes. Assume that $a = 3$, $b = 7$, and $x = 5$.

1. $4a + b \boxed{} 2x + 9$ **2.** $52 - bx \boxed{} ax$ **3.** $a + b + x \boxed{} x^2$

Find the value of the unknown letter in each of the following equations. Look carefully at each equation to determine what you need to do to get the unknown letter alone. Check each solution.

4. $r - 42 = 16$ **5.** $6 = \dfrac{n}{32}$ **6.** $93 = r + 55$

7. $6w = 100 + 2$ **8.** $3(60) + 5 = z - 56$ **9.** $120 = 24t$

10. $10 + 5 = \dfrac{t}{300}$ **11.** $a + 38 = 8^2 - 8$ **12.** $8a = 112$

13. $\dfrac{g}{2} - 11 = 3$ **14.** $0.2x + 20 = 46$ **15.** $1.5h + 0.3 = 1.5$

Answers

1. $=$ **2.** $\neq$ **3.** $\neq$ **4.** $r = 58$ **5.** $n = 192$
6. $r = 38$ **7.** $w = 17$ **8.** $z = 241$ **9.** $t = 5$ **10.** $t = 4500$
11. $a = 18$ **12.** $a = 14$ **13.** $g = 28$ **14.** $x = 130$ **15.** $h = 0.8$

Unit 17 Test

Use a separate piece of paper to work out each problem.

Find the value of each of the following expressions, given that $c = 7$, $r = 2$, and $w = 8$.

1. $4 + 3c - w$

2. $w^2 + 2w + r$

3. $(9c - r)w$

4. $\dfrac{3r + cr + w}{2w - r}$

Put either an $=$ sign or a $\neq$ sign in the boxes between the expressions. Assume that $a = 14$, $b = 6$, and $c = 11$.

5. $(a - c)^2 \,\square\, b$

6. $3b + 7 \,\square\, 2a$

7. $a + b + c \,\square\, a + 2b$

8. $\dfrac{3c + 7}{a + b} \,\square\, b^2$

Find the value of the unknown letter in each of the following expressions. Check each solution.

9. $79 = z + 22$

10. $v - 38 = \dfrac{303}{2 + 1}$

11. $98 = j - 59$

12. $\dfrac{d}{4} = 3^3$

13. $10 - 1 = \dfrac{y}{200}$

14. $k + 64 = (33)(3)$

15. $3y = 42$

16. $8x = (340)(2)$

17. $\dfrac{h}{16} - 2 = 0$

18. $0.02h - 0.8 = 11$

19. $12c - 4 = 20$

20. $\dfrac{s}{8} + 2\dfrac{1}{4} = 4$

Your Answers

1. _____
2. _____
3. _____
4. _____
5. _____
6. _____
7. _____
8. _____
9. _____
10. _____
11. _____
12. _____
13. _____
14. _____
15. _____
16. _____
17. _____
18. _____
19. _____
20. _____

Algebra Review Test

This test covers the material from the last two units on PREPARING FOR ALGEBRA.

Find the value of each of the following expressions.

1. $4^3 + 3^2 \times 2^3 - 2^2$

2. $8^2 \div (4 - 2) + 16$

3. $3(48 - 46)^3$

4. $(8 - 1)^2 - (5 - 2)^3$

5. $(6 + 3^2)^2$

6. $(4 + 3^2)^2 - 6$

7. $\dfrac{\frac{1}{2}}{\frac{1}{2}}$

8. $\dfrac{\frac{1}{2} + \frac{1}{3}}{\frac{1}{6}}$

Find the value of the unknown letter in each of the following equations. Check each solution.

9. $16 = x - 20$

10. $\dfrac{n}{2} = 1 + 3$

11. $d + 8 = 15$

12. $22 = f - 10$

13. $(3)(4) = 4y$

14. $\dfrac{b}{3} = 4 \div 2$

15. $9p = 90 + 3^2$

16. $\dfrac{100 + 152}{2} = 3z$

17. $\dfrac{h}{0.4} + 2 = 11$

18. $\dfrac{k}{30} + 62 = 65$

19. $5c + 8 = 13$

20. $\dfrac{n}{6} + 13 = 15$

Your Answers

1. _____
2. _____
3. _____
4. _____
5. _____
6. _____
7. _____
8. _____
9. _____
10. _____
11. _____
12. _____
13. _____
14. _____
15. _____
16. _____
17. _____
18. _____
19. _____
20. _____

18 | Positive and Negative Numbers

Aims you toward:

- Knowing how to use positive and negative numbers.

- Knowing how to work with "unknowns."

- Adding positive and negative numbers.

- Finding differences between positive and negative numbers (subtracting).

In this unit you will work exercises like these:

- $+4 + (-2)$

- $+3 - (-5)$

- $-5 + (-1) - (2)$

- $-4x + 5x + (-2x)$

Positive and Negative Numbers

Imagine that you have NO money.

We've all been like that at times.

It sounds strange, but you can also have LESS than no money.

Example Suppose you have $600. Since this is money you *have*, let's call it POSITIVE or +.

Now suppose a sweet-talking salesman talks you into buying a used car for only $2,000.

The $600 becomes a down payment, and you owe the balance of

$2,000 − $600 = $1,400.

Since the $1,400 is money you *owe*, let's call it NEGATIVE or −.

Let's take another look at that:

− $2,000	Money you owe
+ $ 600	Money you have
− $1,400	Money you still owe.

(From now on, we'll skip the dollar signs and use only the + and −.)

Example You want to buy a pair of shoes for $60. You have only $40 in your bank account. However, you know that if you write a check for $60, your banker will cover it. Let's see how all of this stacks up.

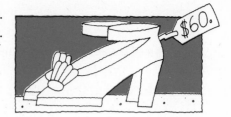

$$-60 \quad \text{Money you owe altogether}$$
$$+40 \quad \text{Money to the good in the bank}$$
$$-20 \quad \text{An overdraft at the bank, or money you still owe}$$

This also can be written as

$$-60 + 40 = -20.$$

Whenever you add positive and negative numbers, think of money. Positive numbers are like money that you have, and negative numbers are like money that you owe. Or, positive numbers are like money that you gain, and negative numbers are like money you lose.

Another Example Marge owes $15 to the telephone company and $35 to the gas company.

$$-15 \quad \text{Money that Marge owes the telephone company}$$
$$-35 \quad \text{Money she owes the gas company}$$
$$-50 \quad \text{Total amount of money owed}$$

We can write this as

$$-15 + (-35) = -50.$$

Notice that we use parentheses around -35 because the $-$ follows the $+$. Otherwise it would be too confusing.

Example Alex received a $7 refund from his credit union. That same day, he received $4 in interest from his bank.

$$+\ 7 \quad \text{Money Alex received, or money to the good}$$
$$+\ 4 \quad \text{More money in his favor}$$
$$+11 \quad \text{Money that Alex now has}$$

This also can be written as

$$+7 + 4 = +11$$

or

$$7 + 4 = \quad 11.$$
$$\uparrow \qquad\qquad \uparrow$$

When no sign appears in front of a number, the number is always positive ($+$).

Exercises

Try some of these. Several problems are worked for you to get you started.

1. $\begin{array}{r} +5 \\ -6 \\ \hline -1 \end{array}$

2. $+5 + (-6) = -1$

3. $\begin{array}{r} -1 \\ -4 \\ \hline -5 \end{array}$

4. $-1 + (-4) = -5$

5. $2 + (-9) =$

6. $\begin{array}{r} 2 \\ -9 \\ \hline \end{array}$

7. $\begin{array}{r} -10 \\ +\ 4 \\ \hline \end{array}$

8. $-10 + 4 =$

9. $-2 + 12 =$

10. $\begin{array}{r} -\ 2 \\ +12 \\ \hline \end{array}$

11. $\begin{array}{r} 0 \\ -4 \\ \hline -4 \end{array}$ You have NO money.
You OWE $4.
You still OWE $4.

12. $0 + (-4) =$

13. $4 - 11 =$

14. $\begin{array}{r} 4 \\ -11 \\ \hline \end{array}$

15. $-3 + (-7) =$

16. $\begin{array}{r} -3 \\ -7 \\ \hline \end{array}$

17. $23 + (-41) =$

18. $\begin{array}{r} 23 \\ -41 \\ \hline \end{array}$

Answers

1. −1	**2.** −1	**3.** −5	**4.** −5	**5.** −7	**6.** −7
7. −6	**8.** −6	**9.** +10	**10.** +10	**11.** −4	**12.** −4
13. −7	**14.** −7	**15.** −10	**16.** −10	**17.** −18	**18.** −18

Example   Bob took up a collection to support a retirement home for elderly racehorses. Mary gave him $12. Helen thought she needed the money more than the horses did, so instead of giving, she took $5 from him. Harry gave $7, and Bob lost $3 through a hole in his pocket. We can show these transactions with positive and negative numbers.

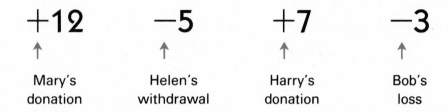

$+12$ -5 $+7$ -3

↑   ↑   ↑   ↑

Mary's   Helen's   Harry's   Bob's
donation   withdrawal   donation   loss

Now let's put them together and find out how much Bob collected.

$$+12 + (-5) + 7 + (-3)$$
$$+7 \quad\;\; + 7 + (-3)$$
$$+14 \quad\;\; + (-3)$$
$$+11 \qquad\qquad \text{or just } \$11$$

Example   Try this problem.

$$+8 + (-4) + (-6)$$
$$+8 + (-4) + (-6)$$
$$+4 \quad\;\; + (-6)$$
$$-2$$

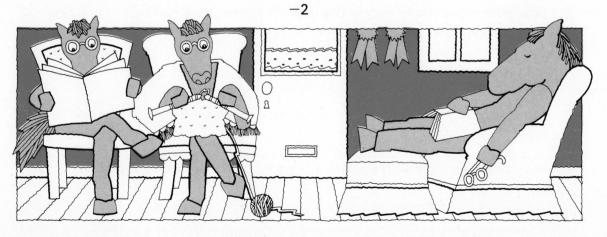

Exercises

Try some exercises. Remember, if you don't see a sign in front of a number, the number is always positive.

1. $6 + (-4) + 2$

$$+2 \quad +2$$

$$+4$$

2. $-8 + 3 + 1$

3. $5 + (-7) + (-4)$

4. $10 + 2 + (-8)$

5. $5 + (-12) + 1$

6. $0 + (-4) + (-6)$

7. $1 + (-9) + 8$

8. $-14 + 7 + (-2) + 1$

9. $-2 + 2 + (-2)$

10. $6 + 0$

11. Herman lost $50 on the dog races, but he made $60 on the horse races. On his way home, he dropped $8 worth of quarters down a manhole. A passerby felt sorry for him and gave him $2. How much did he end up with, if he owed his brother $5?

12. Ken Campbell decided to retrieve used lumber to build his recreation room. He saved $19 in this way. However, his hammer claw broke pulling nails and a new one cost $16. He found $3 in an old envelope between the boards. His electric saw blade was ruined on a hidden nail, and he paid $11 for a new one. He sold leftover boards for firewood and got $5. Use positive and negative numbers to find how much Ken actually saved.

Answers

11. Herman still owes $1 (−$1)   **12.** +$19 − $16 + $3 − $11 + $5 = $0

6. −10   **7.** 0   **8.** −8   **9.** −2   **10.** 6 or +6

1. 4 or +4   **2.** −4   **3.** −6   **4.** 4 or +4   **5.** −6

Working with Unknowns

When we're adding positive and negative numbers, sometimes we have to work with letters of the alphabet called UNKNOWNS. Don't let this worry you. These problems are done exactly the same way as problems you have worked before.

Example $-6a$
             $+5a$

You can still think of money!

$-6a$ Money owed
$+5a$ Money to the good
$-1a$ Money still owed

What is "*a*"? Right now, "*a*" is just coming along for the ride. As long as all the letters (unknowns) in a problem are the same, they can just tag along behind the numbers.

Look at the answer again.

$$-1a$$

The "1" is understood, so you can either put it in ($-1a$) or leave it out ($-a$).

In the same way, the word "apple" means 1 apple. The "1" is understood. So the letter "*a*" means "$1a$."

Let's look at another unknown.

$$r$$

Two things are understood:

1. There's a "1" in front of the *r*.

2. There's a + sign in front of the *r* also.

So "*r*" really means "$+1r$."

Watch out for problems that have DIFFERENT unknowns:

$$2a + 3b \qquad 5y + 7y^2 \qquad 8x + (-2xy).$$

When the unknowns are different, they *can't* just go along for the ride. In another math course you may learn how to work these problems, but when you see different unknowns in this course, just write "can't do it."

Exercises

Try the following problems. Keep thinking of money.

1. $-5m$
   $\underline{2m}$
   $-3m$

2. $5x^2$
   $\underline{-2x^2}$
   $+3x^2$

   Don't worry about unknowns with the same exponents. Right now, the exponents are just going along for the ride.

3. $-t^3$
   $\underline{+2t^3}$

4. $-12y$
   $\underline{+4y^2}$

5. $-12y + 4y$

6. $+3x^4$
   $\underline{-4x^4}$

7. $3x^4 + (-4x^4)$

8. $5abc + (-7abc)$

9. $-xy + 5xy$

10. $8x^2 + (-10x^2)$

Answers

1. $-3m$
2. $+3x^2$
3. t^3 or $+1t^3$
4. Can't do it. The unknowns must be *exactly* alike, with the same exponents.
5. $-8y$
6. $-1x^4$ or $-x^4$
7. $-1x^4$ or $-x^4$
8. $-2abc$
9. $+4xy$ or $4xy$
10. $-2x^2$

Subtracting Positive and Negative Numbers

Earlier in this unit you learned how to ADD positive and negative numbers. Now you will learn how to find the difference between positive and negative numbers. To find differences, you have to SUBTRACT.

Money can be used to understand the difference between two numbers. Remember, if there is a plus sign (+) in front of a number, that number is like money that you have, or money that someone gives you. If there is a minus sign, the number is like money that you owe, or money that you lose.

Example +5 Money to your credit

−5 Money you lose

Example **−2** — **(+3)**

You owe 2 dollars to someone.

You had 3 dollars (+3), but the minus sign in front of it means that you lost the money. (Maybe the 3 dollars fell down a manhole or something!)

−2 **−3**

Money owed Money lost

The answer is −5. This means that you are down 5 dollars altogether.

The difference between −2 and +3 is −5.

Another Example Subtract (find the difference between) −3 and −1.

−3 subtract −1

−3 − (−1)

You owe 3 dollars.

You did owe 1 dollar (−1), but the extra minus sign means that your debt has been taken away.

−3 +1

You still owe the 3 dollars.

When a debt has been taken away, it's like money in the bank. You can call it +1.

The answer is −2. Unfortunately, you are still 2 dollars in the hole.

Example   Find the difference between +1 and +2 (subtract).

+1   subtract   +2

+1   —   (+2)

You have 1   You had 2 dollars in your favor, but the
dollar in your   minus sign in front means that you lost it
favor.   somehow.

+1   —2

You still have   The 2 dollars was lost, so we called
1 dollar.   it —2.

The answer is —1. You are still 1 dollar down on your luck.

The next example will let you make some money for a change!

Example   Find the difference between (subtract) +2 and —4.

+2   subtract   —4

+2   —   (—4)

You have 2   You owed someone 4 dollars (—4) but
dollars to your   the extra minus sign means the debt has
credit.   been canceled.

+2   +4

2 dollars to   Since you no longer owe the 4 dollars, it
your credit.   is money in your favor (+4).

The difference between +2 and —4 is +6. At last you are making
some money!

Example   —1 — (+1)

—1   —1   You owe 1 dollar (—1). You had another dollar (+1),
              but you lost it.

      —2   Since you owed a dollar and then lost another dollar,
              you are now down 2 dollars.

It's time for some exercises on finding differences.

Exercises

Try these. Find the difference (subtract). The first two on each page are done for you.

1. $-1 - (+2)$

$-1 - 2$

-3

2. $+1$
$\underline{-2}$

This problem also can be written as

$+1 - (-2)$

$+1 + 2$

$+3$

3. $-1 - (-2)$

4. $+1$
$\underline{-1}$

5. $-1 - (+1)$

6. $+1 - (+1)$

7. $+3 - (+2)$

8. $-3 - (2)$

9. -3
$\underline{-2}$

10. -1
$\underline{-3}$

11. $-3 - (-1)$

12. $-3 - (-3)$

13. $+1 - (-3)$

14. -1
$\underline{+3}$

15. $+1 - (+3)$

Answers

1. −3	**2.** +3	**3.** +1	**4.** +2	**5.** −2
6. 0	**7.** +1	**8.** −5	**9.** −1	**10.** +2
11. −2	**12.** 0	**13.** +4	**14.** −4	**15.** −2

16. +4
        −3
       ―――
        +7

17. (+4) − (−3) = +7

18. −3
        −4
       ―――

19. −3 − (−4)

20. +3
        +4
       ―――

21. +3 − (+4)

22. +3
        −3
       ―――

23. (+3) − (−3)

24. −1
        −3
       ―――

25. Al owed his dad $600 for a car he purchased. Since Al was going back to school, his dad said that he would drop $200 off the bill for clothes and books. Set up your solution using positive and negative numbers to show how much was left for Al to pay.

26. Heather owed her mother $35 for a phone bill. Since she cleaned the whole house, her mother said that she would take $15 off the bill. Show what Heather had left to pay by setting up the solution using positive and negative numbers.

Answers

16. +7 **17.** +7 **18.** +1 **19.** +1 **20.** −1
21. −1 **22.** +6 **23.** +6 **24.** +2
25. −600 − (−200) = −400 **26.** −35 − (−15) = −20

Exercises

Find the difference (subtract) in each of the following. Remember, the letters beside the numbers just go along for the ride at this point. Don't change them.

1. $-2x - (+4x)$

2. $+3a^2$
   $\underline{-2a^2}$

3. $-2m - (-3m)$

What sign?
↓

4. $2abc$
   $\underline{-4abc}$

5. $-m^3 - (+4m^3)$
   ↑
   (What number?)

6. $4x^2 - (x^2)$

7. $3x^2t - (-2x^2t)$

8. $-2x^2 - (-4x)$

(Watch this one!)

9. $-3abc - (+abc)$

10. $-5y - (-5y)$

11. $xy - (2xy)$

12. $3mt - (4mt)$

13. $x - (x)$

14. $3t - (7m)$

15. $16r - (-7r)$

Answers

13. 0	**14.** Can't do it.	**15.** 23r
10. 0	**11.** $-1xy$ or $-xy$	**12.** $-1mt$ or $-mt$
7. $5x^2t$	**8.** Can't do it.	**9.** $-4abc$
4. 6abc	**5.** $-5m^3$	**6.** $3x^2$
1. $-6x$	**2.** $5a^2$	**3.** 1m or m

Exercises

Here's some more practice in finding differences. Subtract each of the following.

What sign?
↓

1. $5 - (-2)$

2. $5 - (+2)$

3. $-4 - (4)$

4. $-2m - (-2m)$

5. $x^2 - (2x^2)$

6. $3abc - (-1abc)$

7. $xy - (-xy)$

8. $2m - (4m)$

9. $-3x - (-2x)$

10. $-4t^3 - (2t^2)$
(Careful!)

11. $-4ab - (2ab)$

12. $16xy - (7xy)$

13. $abc - (abc)$

14. $rt - (-3rt)$

15. $-2xt^2 - (2x^2t)$

Answers

1. $+7$

2. $+3$

3. -8

4. 0

5. $-1x^2$ or $-x^2$

6. $+4abc$

7. $+2xy$

8. $-2m$

9. $-x$ or $-1x$

10. Can't do it.

11. $-6ab$

12. $+9xy$

13. 0

14. $+4rt$

15. Can't do it.

Review: Positive and Negative Numbers

At the beginning of this unit, you learned to add positive and negative numbers by thinking of money. Positive numbers are like money you gain, and negative numbers are like money you owe.

Next, you learned to find the differences between positive and negative numbers by thinking of money again. Finding the difference between two numbers is really subtracting.

Here is an example of combining $+$ and $-$ in one problem.

Example $\quad (+4) + (-5) - (-2)$ You have 4 dollars ($+4$), and you are going to add a debt of 5 dollars (-5).

$\qquad\qquad -1 \qquad -(-2)$ This leaves you 1 dollar (-1) in the hole. You also have a debt of 2 dollars that you are going to remove. This is the same as 2.

$\qquad\qquad -1 \qquad +2$ You still owe 1 dollar, but you also have 2 dollars in your favor.

$\qquad\qquad\qquad +1$ The answer is $+1$, or 1.

Example $\quad (-2) - (+3) + (-1)$

$\qquad\qquad -5 \qquad + (-1)$

$\qquad\qquad\qquad -6$

Example $\quad (2) + (-7) - (-3) + (-1)$

$\qquad\qquad -5 \qquad - (-3) + (-1)$

$\qquad\qquad\qquad -2 \qquad + (-1)$

$\qquad\qquad\qquad\qquad -3$

Exercises

Here's an exercise set with only six problems. Isn't that nice?

1. $(-1) + (-2) - (-3)$

2. $(3) - (2) + (-1)$

       ↑   ↑

     What sign?

3. $(-2) + (-3) - (1)$

4. $(4) - (-2) - (1)$

5. $(3) - (3) - (4)$

6. $(-5) + (3) + (-4)$

Answers

1. 0   2. 0   3. −6   4. 5   5. −4   6. −6

Exercises

Try these. This time, some unknowns (letters) are mixed in with the numbers.

1. $(-5) + (-1) - (-3)$

2. $(1) - (-2) + (-3)$

3. $(4y) + (-2y) - (-y)$

4. $(-xy) - (2xy) + (3xy)$

5. $(3x^2) - (x^2) + (2x^2)$

6. $(-4m) - (m) + (5m)$

7. $(2) - (-5) + (-3)$

8. $(-mt^2) - (-3mt^2) + (-2mt^2)$

9. $(4) - (-9) + (2) - (1)$

10. $(-5xy) + (-2y) - (7m)$

Answers

1. -3 **2.** 0 **3.** $3y$ **4.** 0 **5.** $4x^2$ **6.** 0 **7.** 4 **8.** 0 **9.** 14 **10.** Can't do it.

Exercises

Simplify the following.

1. $(3) + (-2) - (1)$

2. $(-3) - (-1) - (-1)$

3. $(-11) + (4) + (-7)$

4. $(8) - (-2)$

5. $(4x) - (-2x) + (3x) - (x)$

6. $(-m) + (2m) - (-3m) + (m)$

7. $(2ab) - (-3ab) - (2ab)$

8. $(-64) - (14) + (-31)$

There is a shortcut for doing all the problems, but you have to know how to multiply positive and negative numbers FIRST.

So for now, keep thinking of money, and ask your teacher about the shortcut.

Answers

5. $8x$ **6.** $5m$ **7.** $3ab$ **8.** -109
1. 0 **2.** -1 **3.** -14 **4.** 10

Review Exercises

Here's a final review before your test. Simplify the following.

1. $(27) + (-16) - (22)$

2. $-(-82)$

$0 - (-82)$

↑

You can put a zero here.

3. $-(64)$

4. $(-18) - (12) + (-15)$

5. $-(19) + (-17)$

6. $(-13) - (-41) + (25)$

7. $(24mt) - (mt)$

8. $(-11tm) + (6tm)$

9. $15mt^2 - (4mt)$

Watch it!

10. $(-29tm^2) + (8tm^2)$

11. $-13 + (-3) - (-1)$

12. $(18 - 7) - (5 + 11)$

13. $50 - (6 + 9) + (-5) - (-30)$

14. $-(-8) + (7) + (-4) - (-2)$

Answers

1. −11	**2.** 82	**3.** −64	**4.** −45	**5.** −36
6. 53	**7.** 23mt	**8.** −5tm	**9.** Can't do it.	
10. −21tm²	**11.** −15	**12.** −5	**13.** 60	**14.** 13

Unit 18 Test

Use a separate piece of paper to work out each problem. Simplify the problems below. Think of money.

1. $-4 + 3$ **2.** $-4 - (3)$ **3.** $5 - (-2)$

4. $(-1) + (2)$ **5.** $-2r + (-4r)$ **6.** $-t - (3t)$

7. $(3) + (-2) - (-3)$ **8.** $(-1) - (1) + (-2)$

9. $(4) - (2) - (-1)$ **10.** $(-6) + (-2) + (1)$

11. $(-42) + (-30) - (-12)$

12. $(61) - (20) - (-14)$

13. $(36x^2) - (-10x)$ **14.** $-(-18)$

15. $(3a) - (-12a)$

16. $(-16xy) + (-2xy) - (xy)$

17. $22 - (9 - 6) - (4 + 9)$

18. $-13 + (24 - 19) - (-8)$

19. Suppose you have $400 in the bank, and buy a new home computer for $1,500. Set this up using positive and negative numbers to show how much you still owe.

20. A week after you bought the computer, the store calls to tell you that a sale discount had been overlooked. Using your answer from problem 19, show a debt of $150 being removed from what you owe.

Your Answers

1. _____
2. _____
3. _____
4. _____
5. _____
6. _____
7. _____
8. _____
9. _____
10. _____
11. _____
12. _____
13. _____
14. _____
15. _____
16. _____
17. _____
18. _____
19. _____
20. _____

Test Answers

Unit 1

1. 175	**2.** 1,067	**3.** 1,458	**4.** 1,247	**5.** 17,941
6. 1,179	**7.** 370	**8.** 976	**9.** 220	**10.** 2,650
11. 2,534	**12.** 637	**13.** $3,570	**14.** 1,260,000	
15. 3,425,348	**16.** 832,605,000	**17.** 9,131,154		

18. Two hundred fifty thousand

19. Five million, one hundred forty-six thousand, two hundred ninety-eight

20. One hundred ninety-eight thousand, six hundred twelve

Unit 2

1. 2,121	**2.** 4,532	**3.** 56	**4.** 4,888
5. 38,882	**6.** 4,611	**7.** 24,396	**8.** 27,321
9. 20,239	**10.** 80,647	**11.** 311	**12.** 60
13. 4	**14.** 343	**15.** 5,336	**16.** 4,606
17. 8,886	**18.** 23,472	**19.** $39,967.33	**20.** 1,717 pages

Unit 3

1. 24	**2.** 42	**3.** 1,263	**4.** 576	**5.** 595
6. 427	**7.** 336	**8.** 810	**9.** 3,392	**10.** 4,982
11. 3,240	**12.** 1,200	**13.** 3,234	**14.** 289,984	
15. 14,652	**16.** 221,238	**17.** 18,040,064	**18.** 32,496	
19. 571,250	**20.** 645,018	**21.** 204,592	**22.** 17,500	
23. 1,730,026	**24.** 122,171	**25.** $4,212	**26.** $13,860	

Unit 4

1. 2 R21	**2.** 2 R264	**3.** 4 R26	**4.** 4,020 R12	**5.** 68 R1
6. 16 R125	**7.** 7,110	**8.** 10 R15	**9.** 344 R18	**10.** 75 R78
11. 130 R4	**12.** 114 R58	**13.** 428 R8	**14.** 39 R46	
15. 8,009 R5	**16.** 2 R220	**17.** 73 times	**18.** 25 minutes	

Review Test

1. 37,728 **2.** 346 **3.** 357 **4.** 1,312 **5.** 153

6. 293 R19 **7.** 729 **8.** 4,527 **9.** 54 **10.** 49

11. 141 R9 **12.** 2,334,000 **13.** $75

14. 3 sweaters, $15 left

Unit 5

1. $\frac{23}{5}$ **2.** $2\frac{1}{5}$ **3.** $\frac{8}{3}$ **4.** $\frac{17}{5}$ **5.** $\frac{33}{7}$

6. $1\frac{5}{7}$ **7.** $20\frac{5}{8}$ **8.** $3\frac{1}{18}$ **9.** 8 **10.** 18

11. $\frac{15}{32}$ **12.** $\frac{3}{7}$ **13.** $\frac{4}{13}$ **14.** $2\frac{8}{11}$ **15.** $4\frac{2}{5}$

16. $\frac{2}{25}$ **17.** $3\frac{5}{9}$ **18.** $1\frac{1}{6}$ **19.** $\frac{9}{40}$ **20.** $375

21. $642 \times \frac{2}{3} = 428$ employees

22. $599\frac{1}{3} \div 4 = 149\frac{5}{6}$ cubic yards

Unit 6

1. $1\frac{1}{12}$ **2.** $1\frac{3}{40}$ **3.** $5\frac{8}{15}$ **4.** $9\frac{19}{24}$ **5.** $24\frac{5}{12}$ **6.** $11\frac{5}{28}$

7. $15\frac{17}{20}$ **8.** $10\frac{1}{24}$ **9.** $1\frac{5}{24}$ **10.** $\frac{1}{2}$ **11.** $96\frac{5}{8}$ **12.** $11\frac{1}{4}$

13. $11\frac{1}{2}$ **14.** $14\frac{1}{8}$ **15.** $1\frac{3}{10}$ **16.** $1\frac{23}{24}$ **17.** $2\frac{25}{36}$ **18.** $5\frac{29}{35}$

19. $32\frac{7}{24}$ hours **20.** $\frac{29}{32}, \frac{7}{8}, \frac{13}{16}, \frac{3}{4}$

21. $\frac{9}{10} + \frac{9}{10} = \frac{18}{10} = 1\frac{8}{10} = 1\frac{4}{5}$

Review Test

1. $\dfrac{2}{21}$ 2. $\dfrac{5}{24}$ 3. $\dfrac{5}{12}$ 4. $\dfrac{9}{16}$ 5. $\dfrac{7}{8}$ 6. $4\dfrac{19}{32}$

7. $1\dfrac{1}{2}$ 8. $4\dfrac{3}{8}$ 9. $3\dfrac{13}{20}$ 10. $6\dfrac{13}{22}$ 11. $31\dfrac{1}{9}$ 12. $2\dfrac{6}{49}$

13. $4\dfrac{11}{21}$ 14. $\dfrac{6}{7}$ 15. $16\dfrac{10}{11}$ 16. $\dfrac{20}{21}$

17. $4\dfrac{1}{4} - 3\dfrac{7}{8} = \dfrac{3}{8}$ of a pound   18. $\dfrac{1}{4} + \dfrac{3}{8} = \dfrac{5}{8}$ of the store

19. $\dfrac{2}{5} \times \dfrac{2}{3} = \dfrac{4}{15}$ of the total   20. $37\dfrac{1}{2} \div 5 = 7\dfrac{1}{2}$ hours

Unit 7

1. Hundreds   2. Hundredths   3. Hundred-thousandths
4. Tenths   5. Ten-thousandths   6. Thousandths
7. Forty-five and three hundred fifty-four thousandths
8. Sixteen and seventy-five hundredths
9. 19.8   10. 50.097   11. 95.313   12. 12.028   13. 127.55
14. $18.51   15. 57.63   16. 4.03   17. 117.4
18. 26.5 miles   19. $8.37   20. 1,139.989 meters

Unit 8

1. 166.32   2. 4.14   3. 4.173   4. 164.19   5. 3.685
6. 31.286   7. 442.68   8. 89.352   9. 0.529   10. 1.918
11. 0.134   12. 1.585   13. 9.353   14. 14.683   15. 47.259
16. 0.002   17. 9.881   18. 59.492   19. $1,814.82
20. 21.24 miles per gallon   21. 0.556 revolutions per second
22. 3,575 feet

Review Test

1. Ones **2.** Hundredths **3.** Tenths

4. Thousands **5.** Ten-thousandths **6.** Tens

7. 51 or 51.0 **8.** 3.62 **9.** 15.901 **10.** 20.82

11. $3.22 **12.** $44.15 **13.** 21.016 **14.** 0.862

15. 1,525 **16.** 5,060.02 **17.** 0.108504 **18.** 2.6

19. 0.3125 **20.** 0 **21.** 6 or 6.000 **22.** $4.93

Unit 9

1. $3\frac{4}{25}$, 3.16 **2.** $\frac{3}{1}$, 300% **3.** 2.25, 225%

4. 2.125, 212.5% **5.** $5\frac{1}{250}$, 500.4% **6.** $\frac{3}{100}$, 0.03

7. $\frac{1}{1}$, 1 **8.** 1.273, 127.3% **9.** $\frac{4}{25}$, 0.16

10. $\frac{4}{1}$, 400% **11.** $2\frac{19}{50}$, 2.38 **12.** 0.778, 77.8%

13. $\frac{147}{200}$, 73.5% **14.** 2.83, 283% **15.** $4\frac{1}{40}$, 402.5% **16.** $\frac{1}{100}$, 0.01

17. 78% **18.** $\frac{3}{8} = 37.5\%$ **19.** 0.06 = 6% **20.** $12\% = \frac{3}{25}$

Unit 10

1. Yes **2.** No **3.** Yes **4.** Yes **5.** No **6.** Yes

7. $K = 28$ **2.** $F = 2.7$ **9.** $K = 35$ **10.** $B = 3$

11. $d = 36$ **12.** $b = 28$ **13.** $a = 21$ **14.** $g = 25$

15. $k = 28$ **16.** $e = 10.8$ **17.** $r = 10$ **18.** 72 feet

19. 13.5 millimeters **20.** $16,000,000

Unit 11

1. 20%   2. 52   3. 76   4. 5.6   5. 320

6. 26%   7. 75   8. 50   9. 84   10. 200%

11. 10   12. 25%   13. $110.66   14. $22,000

15. $20,000   16. $35.40   17. 6%   18. $760.45

Review Test

1. 0.1875, 18.75%   2. 47.5%, $\frac{19}{40}$ 3. 0.68, $\frac{17}{25}$

4. 208%, $2\frac{2}{25}$ 5. $3\frac{4}{25}$, 3.16   6. 2.8, 280%

7. Yes   8. Yes   9. No   10. 4

11. 10.5   12. 162   13. 64   14. 369

15. 20%   16. 50   17. $3,347.50

18. 37.5%, $\frac{3}{8}$ 19. 16.6%   20. 700 pounds

Unit 12

1. $\frac{1}{10}$ 2. $\frac{1}{100}$ 3. 100   4. $\frac{1}{1000}$ 5. km

6. hm   7. dam   8. dm   9. cm   10. mm

11. 5,000   12. 0.005   13. 0.018   14. 2,000   15. kl

16. hl   17. dal   18. dl   19. cl   20. ml

21. 24,000,000   22. 0.0032   23. kg   24. hg

25. dag   26. dg   27. cg   28. mg   29. 645

30. 0.008   31. 80   32. 0.000001   33. $\frac{1}{2}$ 34. 42

35. About 12 miles per hour

Unit 13

1. Parallelogram, 13 meters
2. Quadrilateral, 37 cm
3. Trapezoid, 7.50 miles
4. Square, 36 km
5. 4000 mm^2
6. 2268 cm^2
7. 880 feet^2
8. 78 mm^2
9. 110 inches^2
10. 6.50 m^2
11. 96 inches^3
12. 60 cm^3
13. 49 inches^3
14. 27 feet^3
15. 2100 mm^3
16. 392 cm^3

Unit 14

1. 81.64 cm, 530.66 cm^2
2. 37.68 mm, 113.04 mm^2
3. 31.40 inches, 78.50 inches^2
4. 37.68 mm, 113.04 mm^2
5. 28.27 cm, 47.49 cm^2
6. 375.22 mm, 8366.53 mm^2
7. Yes. (Area of big pizza = 200.96. Area of both small pizzas together = 100.48.
8. $1,538.60 \text{ inches}^3$
9. $5,086.80 \text{ m}^3$
10. 981.25 cm^3

Review Test

1. Parallelogram, 32 inches, 36 inches^2
2. Rectangle, 17 cm, 16.5 cm^2
3. Trapezoid, 11.5 feet, 7 feet^2
4. Circle, 50.24 m, 200.96 m
5. Triangle, 14.5 mm, 9 mm^2
6. Triangle, 44 cm, 98 cm^2
7. 23.13 m, 31.79 m^2
8. 42 mm, 96 mm^2
9. 24 m, 32 m^2
10. 76 cm, 330 cm^2
11. 100.48 inches^3
12. 6231.33 cm^3
13. $1,271,700 \text{ m}^3$
14. 12.56 inches^3

Unit 15

1. 2
2. $\frac{1}{12}$
3. $\frac{7}{18}$
4. 0.156
5. $\frac{1}{12}$
6. 47.12
7. 8.24
8. $\frac{1}{4}$
9. 35.25
10. $\frac{5}{9}$
11. 2.515
12. 86.4 seconds; or 1 minute, 26.4 seconds
13. $60.90
14. 492 tickets

Unit 16

1. 60   **2.** 20   **3.** 0   **4.** 51   **5.** 196   **6.** 55

7. 28   **8.** 186   **9.** 1,215   **10.** 56   **11.** 40   **12.** 150

13. 810   **14.** 32   **15.** 0   **16.** 180   **17.** 5

18. 7   **19.** $\frac{1}{2}$ **20.** 1   **21.** $\frac{3}{16}$ **22.** 10

Unit 17

1. 17   **2.** 82   **3.** 488   **4.** 2   **5.** $\neq$

6. $\neq$ **7.** $\neq$ **8.** $\neq$ **9.** $z = 57$ **10.** $v = 139$

11. $j = 157$ **12.** $d = 108$ **13.** $y = 1800$ **14.** $k = 35$ **15.** $y = 14$

16. $x = 85$ **17.** $h = 32$ **18.** $h = 590$ **19.** $c = 2$ **20.** $s = 14$

Review Test

1. 132   **2.** 48   **3.** 24   **4.** 22   **5.** 225

6. 163   **7.** 1   **8.** 5   **9.** $x = 36$ **10.** $n = 8$

11. $d = 7$ **12.** $f = 32$ **13.** $y = 3$ **14.** $b = 6$ **15.** $p = 11$

16. $z = 42$ **17.** $h = 3.6$ **18.** $k = 90$ **19.** $c = 1$ **20.** $n = 12$

Unit 18

1. -1 **2.** -7 **3.** 7   **4.** 1   **5.** $-6r$ **6.** $-4t$

7. 4   **8.** -4 **9.** 3   **10.** -7 **11.** -60 **12.** 55

13. Can't be done.   **14.** 18   **15.** $15a$

16. $-19xy$ **17.** 6   **18.** 0

19. $\$400 + (-\$1,500) = -\$1,100$ **20.** $-\$1,100 - (-\$150) = -\$950$

Index